The Rise of Biodesign

Contemporary Research Methodologies for Nature-Inspired Design in China

Edited by Mary Polites

Tongji University Press

Executive Editor: Crisie Yuan
Copy Editor: Mark Zuiderveld
Book Design: Mary Polites, MAPS

National Library of China Cataloguing-in-Publication Data. A catalogue record for this book is available from the National Library of China ISBN 978-7-5608-8189-8

Printed and Bound in the People's Republic of China.

Contents

Preface

Bioinspired design is both simple and complex.

Simple as it provides an easy reference to discuss ideas through a means we all can relate and feel inspired by – *the natural world*. Nature easily engages those from a wide range of backgrounds, from environmental enthusiasts to professional designers and scientists, as well as the general public.

Complexity arises when we need to understand beyond what we see. At many levels, we are still exploring how natural systems are interrelated and how to start mapping these connections. Whether in science research or in design practice, interdisciplinary collaboration is one of the main paths to unlock the complexity of the natural world, and aim to successfully confront today's environmental challenges. Despite this, even nature-inspired fields are not immune to fragmentation, and the reason is well-known: scientific reductionism allows us to move faster. But do we know where we are heading? The biomimicry movement initiated by Janine Benyus two decades ago fused the principles of biomimetics with those of environmental awareness and ecological design. The statement was clear, we should continue learning from nature and aiming for innovation, but with a clear purpose: the preservation of our planet and the reconnection of the human species with the natural world.

In this context, the BiDL (Biomimetic Design Lab) was established as the first Biomimicry lab[1] in China in 2012, as part of the College of Design and Innovation (D&I) at Tongji University in Shanghai. The lab represented a merger between sustainability and contemporary design methods applied in a wide variety of fields, from architecture to industrial, interaction and service design.

---

1 Biomimicry is defined by the Biomimicry Institute as the intended emulation of nature which focuses on the methods and concepts of study revolving around sustainability. BiDL lab fully recognizes other biomimetic labs in China having been established prior and includes them in this book to encompass the whole field of nature inspired design.

The BiDL is an unusual entity in the Chinese education environment, as courses associated with biomimicry or biodesign are typically reserved for master or expert level training. This means that specialized level education is made accessible for younger students, and opened to a variety of design degrees. Additionally, through the core sustainability classes, biomimicry is an integral topic for the undergraduate and graduate lectures.

This book has a double purpose: it celebrates the initial 5 years of BiDL, highlighting the efforts, challenges and achievements in education through a collection of students' work at the D&I, from undergraduate studios to master thesis.

And second, there was a strong interest to expand discussions about bioinspired design beyond the classroom, thus initiating conversations with other Chinese biomimetic labs[2] as a way of exchanging methodologies, sharing ambitions, and building collaborations.

This book offers unique insight into the current state of biomimetics and biomimicry in a challenging but exciting moment for bioinspired fields. Through professional research and students projects, this collection of work shines light on educational methods currently being explored in some of China's top institutions, as part of a national environmental agenda that is critical, not only for the future of the country, but for our entire planet.

2 See Biomimicry Labs Map in pp. 20-25 for comparison.

Acknowledgments

This book represents a collaboration of work made possible due to the interdisciplinary design environment at the College of Design and Innovation, at Tongji University.

I would also like to thank the following individuals for their assistance, coordination and contribution:

The team of students around the world that volunteered their time. Specifically, Jiarui Tan, Yadian Wang and Raquel Paramo.

The team and founders of BiDL – Pius Leuba and Xiaocun Zhu.

Contributors from our fellow Chinese Universities, specifically:
Yan Liu from Jilin University, Yongmei Zheng and Lei Jiang from Beihang University, and Dengteng Ge from Donghua University.

The interviewees who dedicated their time to provide insightful conversations: Pius Leuba, Marie Dariel, Camille Fong, Sherry Ritter, Dengteng Ge, Yongmei Zheng, George Jeronimidis, Evan Greenberg and Julian F. Vincent.

Tongji University Press, namely Crisie Yuan and her team for their support.

To my family, friends, and partner, Ignacio Lopez Buson, for their love and support.

How to Read this Book

Chapter 1 is an extensive collection of perspectives curated from some of the earliest researchers in the field that have made biodesign what it is today. These interviews cover biomimetic design from all areas, biologists, zoologists, architects, and environmental designers. This chapter is for readers looking to discover the nuances of biomimetic design across many disciplines and is suited for practitioners as well as the general reader.

Chapters 2, 3, and 4 include works that provide examples of biomimetic design in practice. These chapters are well suited for researchers or students looking for information on current projects explored in China. Additionally, these texts have been developed so that a general audience can understand even the most scientific process.

Chapter 2 is an overview of the terms and concepts most used in the biomimetic projects found in chapter 3. This includes illustrations that break down patterns and unique aspects found in nature that are assumed by the biomimetic committee and often explored throughout the field.

Chapter 3 shows the research and outcomes by biomimetic and biomimicry labs from Jilin University, Beihang University, Donghua University and Tongji University.

Chapter 4 focuses on the design education methods covered at BiDL lab. These studies aim to show how students are introduced to topics of nature at the urban, human, and microscopic scale. These projects show how ideas were abstracted from natural logics and applied to design applications.

Chapter 5 offers a conclusion and perspective on the future of bioinspired design in China.

The final appendix provides a collection of references and educational tools that students and researchers may find useful.

Introduction

Bioinspiration is a broad field that encompasses many facets of biology, engineering and design. The research and development of materials, systems and techniques based on natural phenomena is considered one of the main contributors for innovation in applied sciences, and a critical driver for future advances in sustainability.

Bioinspiration is anticipated to have a significant contribution from the emerging markets in growing economies, such as India and China, and the byproducts from this field are estimated to greatly help towards the alleviation of world pollution in the next 20 years. However, the inspired-by-nature label does not necessarily imply a sustainable process, a waste-free outcome or an ecologically driven agenda. And that is how the biomimicry movement initiated by Janine Benyus in 1997 stood out among the rest of the biologically inspired fields.

In the last two decades, a growing number of biomimetic labs in China have been focusing on new approaches in the fields of material design, mechanics and engineering based on natural logics. However, the concept of biomimicry is still novel and abstract for many Chinese audiences. In this context, BiDL lab was established as one of the first centers to integrate sustainability, design and innovation within a Chinese university.

Leaf Veins

Detailed view of a leaf

The Rise of Bioinspiration

Although nature has been inspiring humankind in every possible field (science, art, design...) for thousands of years, there has been a cyclical interest in the natural world undeniable linked to historic environmental crisis caused by human activity. Today, due to a general awareness of global warming and its consequences, a renovated environmental concern has emerged since the beginning of the 21st century.

In addition, major technological breakthroughs experienced since the 1990s have allowed for unprecedented improvements in the measurement, analysis and understanding of any natural phenomena, including the simulation and prediction of environmental processes. Consequently, a closer relationship between nature and humankind is "in high demand" today, thus resulting into the widespread of nature-related fields in contemporary culture.

Despite the fact that bioinspiration has been recurrently present in scientific research since the 1950s, it was Janine Benyus' book, *Biomimicry: Innovation Inspired by Nature* (1997), a main trigger for the term to come out of the lab and dive into our daily life. In this regard, she managed to make the biomimicry movement stand out from previously established fields (biomimetics, bionics) by adding environmental awareness and sustainability to the inspired-by-nature equation.

Today, the concept of bioinspired design is developing at an impressive rate, and it is attractive enough to a wide audience. However, the extent to which bioinspiration can influence our lives through its different ramifications is still relatively unknown. In that sense, this book aims to serve as an introduction to the umbrella term of bioinspired design, and bring clarity to the field by explaining the similarities and dissimilarities between its different specializations.

Human Cities

D&I presented six projects that showcased approaches to social design, system design, open design, and strategic design, developing strategies for sustainability innovation.

History of Bioinspired Design

Besides the widespread recognition of Leonardo Da Vinci as the first modern bioinspired designer more than 500 years ago, the contemporary terms related to bioinspiration in design were developed halfway through the 20th century. The term biomimetics was coined by American biophysicist and polymath Otto Schmitt in the 1950s[1]. It was used during his research on the Schmitt trigger, which studied the nerves in squid, attempting to engineer a device that replicated the biological system of nerve impulse propagation.[2] From then on, he continued to develop research focusing on technology that could mimic natural systems. In 1957, what was known as *biophysics* at that time (a branch of science concerned with the application of physical principles and methods to biological problems), was not sufficient in his view, and consequently, he developed the term biomimetics. Today, biomimetics as defined by the Merriam-Webster dictionary as the study of biological materials, structures, functions, mechanisms and processes, especially for the purpose of synthesizing similar products by artificial mechanisms which mimic natural ones.

The other term developed around the same period of time, and often used as synonymous with biomimetics, is bionics. The concept was coined by Jack Steele of the US Air Force in 1960 at a meeting at the Wright-Patterson Air Force Base.[3] It most often suggests the combination of engineering and biology for technical end applications. Bionics appears to have more frequency of use in Europe and China, as shown in publication and lab titles, listed in the appendix. Bionics has been defined as a scientific discipline that "deals with the technical implementation and application of constructional processing and development principles of biological systems".[4]

---

1 Vincent, Julian F.v., Olga A. Bogatyreva, Nikolaj R. Bogatyrev, Adrian Bowyer, and Anja-Karina Pahl. "Biomimetics: Its Practice and Theory." Journal of The Royal Society Interface 3, no. 9 (2006): 471-82. doi:10.1098/rsif.2006.0127.

2 Harkness, Jon M. "In Appreciation: A Lifetime of Connections: Otto Herbert Schmitt, 1913 - 1998." Physics in Perspective (PIP) 4, no. 4 (2002): 456-90. doi:10.1007/s000160200005.

3 Gundlach, Carsten. "Bionics in Patents - Semantic Base Analysis for the Exploitation of Bionic Principles in Patents." In Current Scientific and Industrial Reality: Proceedings of the TRIZ-Future Conference 2007 ; Frankfurt, Germany, November, 6th - 8th, 2007. Kassel: Kassel Univ. Press, 2007.

4 Harkness, Jon M.

On the other hand, biomimicry, coined by Janine Benyus in 1997, is part of a general trend of convergence between biology, engineering, and sustainability.[5] The meaning associated with biomimicry is *good design abstracted from nature*. The definition, however, according to the Merriam-Webster Dictionary, references directly to biomimetics and defines it as the imitation of natural biological designs or processes in engineering or invention. This does not convey all social and environmental implications that reference the sustainable and holistic ethos. In that sense, the Biomimicry Institutes's definition as "an approach to innovation that seeks sustainable solutions to human challenges by emulating nature's time-tested patterns and strategies" more accurately describes the agenda behind the term.

In an attempt to differentiate between the three concepts, we could suggest that:

- biomimetics aims to replicate organic and inorganic structures, logics and processes found in nature, to solve existing problems or to find new applications;

- bionics focuses on the study of biological systems and the generation of artificial counterparts of those entities that could perform in a similar or improved manner for different purposes;

- biomimicry learns from nature to reproduce materials, objects, processes and methods that come together with an idea of aiming towards social, economic and environmental sustainability.

In the context of this book, we will assume that the portions focusing on the scientific aspects of biological design can be considered biomimetics or bionics, while in the areas that focus on sustainability, the term biomimicry will be used. Also, to put some emphasis on the similarities among all the bioinspired design fields, the following time line traces the origin and current state of all related terms through seminal books and key figures in the fields of science, literature, art and design.

5 Yoram, Helfman Cohen Yael. Reich. Biomimetic Design Method for Innovation And Sustainability. S.l.: Springer, 2018.

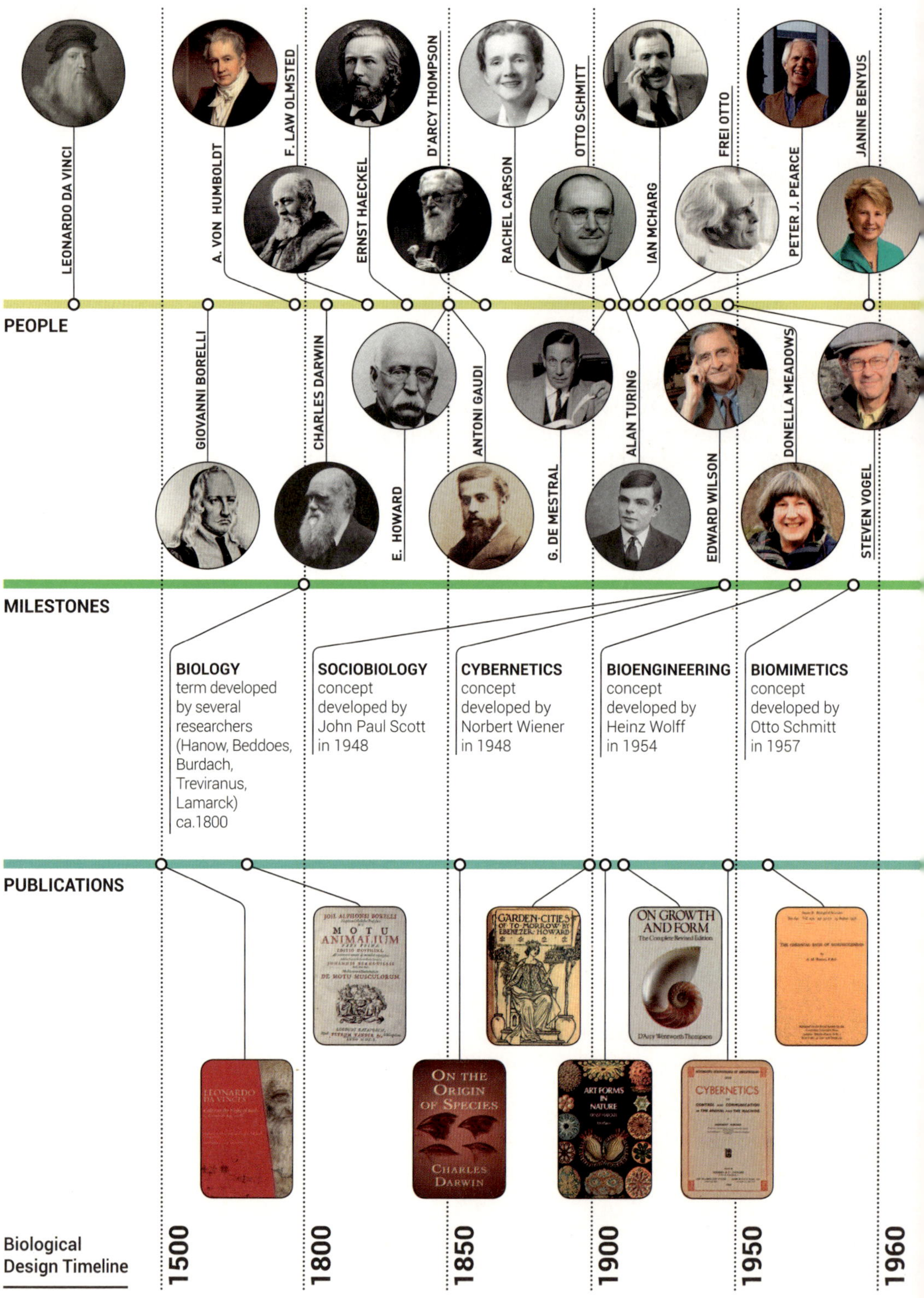
PEOPLE
LEONARDO DA VINCI
A. VON HUMBOLDT
F. LAW OLMSTED
ERNST HAECKEL
D'ARCY THOMPSON
RACHEL CARSON
OTTO SCHMITT
IAN MCHARG
FREI OTTO
PETER J. PEARCE
JANINE BENYUS
GIOVANNI BORELLI
CHARLES DARWIN
E. HOWARD
ANTONI GAUDI
G. DE MESTRAL
ALAN TURING
EDWARD WILSON
DONELLA MEADOWS
STEVEN VOGEL
MILESTONES
BIOLOGY term developed by several researchers (Hanow, Beddoes, Burdach, Treviranus, Lamarck) ca.1800
SOCIOBIOLOGY concept developed by John Paul Scott in 1948
CYBERNETICS concept developed by Norbert Wiener in 1948
BIOENGINEERING concept developed by Heinz Wolff in 1954
BIOMIMETICS concept developed by Otto Schmitt in 1957
PUBLICATIONS
DE MOTU ANIMALIUM
LEONARDO DA VINCI
ON THE ORIGIN OF SPECIES
CHARLES DARWIN
GARDEN CITIES OF TO-MORROW BY EBENEZER HOWARD
ART FORMS IN NATURE
ON GROWTH AND FORM
The Complete Revised Edition
D'Arcy Wentworth Thompson
CYBERNETICS
Biological Design Timeline
1500
1800
1850
1900
1950
1960

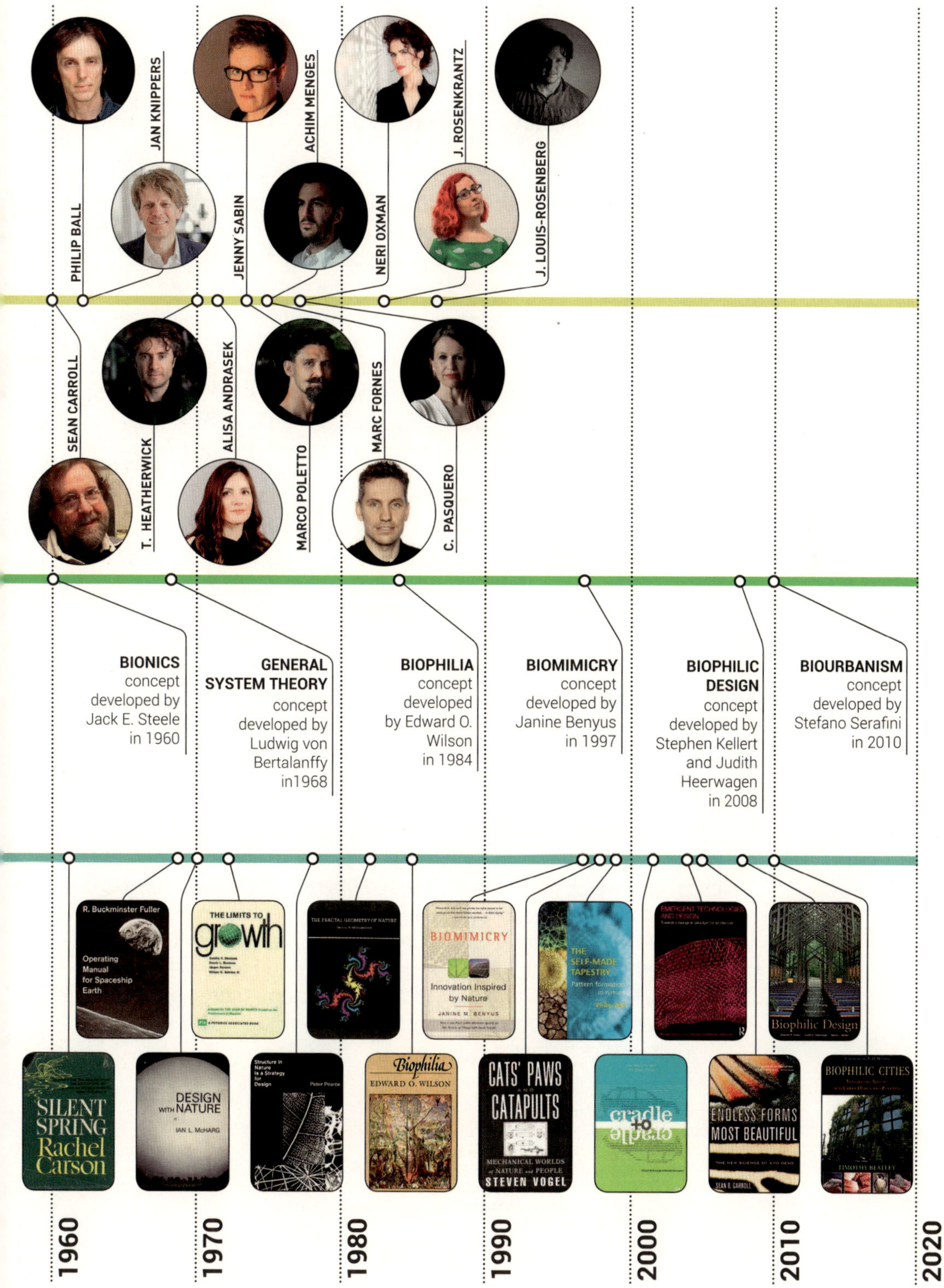
PHILIP BALL
JAN KNIPPERS
JENNY SABIN
ACHIM MENGES
NERI OXMAN
J. ROSENKRANTZ
J. LOUIS-ROSENBERG
SEAN CARROLL
T. HEATHERWICK
ALISA ANDRASEK
MARCO POLETTO
MARC FORNES
C. PASQUERO
BIONICS concept developed by Jack E. Steele in 1960
GENERAL SYSTEM THEORY concept developed by Ludwig von Bertalanffy in1968
BIOPHILIA concept developed by Edward O. Wilson in 1984
BIOMIMICRY concept developed by Janine Benyus in 1997
BIOPHILIC DESIGN concept developed by Stephen Kellert and Judith Heerwagen in 2008
BIOURBANISM concept developed by Stefano Serafini in 2010
R. Buckminster Fuller
Operating Manual for Spaceship Earth
THE LIMITS TO growth
THE FRACTAL GEOMETRY OF NATURE
BIOMIMICRY
Innovation Inspired by Nature
JANINE M. BENYUS
THE SELF-MADE TAPESTRY
Pattern formation in nature
Philip Ball
EMERGENT TECHNOLOGIES AND DESIGN
Biophilic Design
SILENT SPRING
Rachel Carson
DESIGN WITH NATURE
IAN L. McHARG
Structure in Nature Is a Strategy for Design
Peter Pearce
Biophilia
EDWARD O. WILSON
CATS' PAWS AND CATAPULTS
MECHANICAL WORLDS of NATURE and PEOPLE
STEVEN VOGEL
cradle to cradle
ENDLESS FORMS MOST BEAUTIFUL
SEAN B. CARROLL
BIOPHILIC CITIES
TIMOTHY BEATLEY
1960
1970
1980
1990
2000
2010
2020

The Potential Impact of Bioinspired Design

In the sciences, bioinspiration as a topic has been developing since the early 1980s, primarily in Europe, as discussed in the interviews (***see George Jeronimidis***). This interest in biological design has continued to grow to the present, as our tech is pushing boundaries and our environmental consciousness is demanding new solutions. The achievements and explorations of bioinspiration have been touted in numerous articles that span fields such as *Forbes*[6], the *Economist*[7], and the *Financial Times*[8] . Significant reports have additionally suggested how these developments will impact international economies.

"Bioinspiration: An Economic Progress Report", put forth by the Fermanian Business & Economic Institute (FBEI) introduced an extensive analysis of the topic in 2013. According to these studies, by 2025 the field of bioinspiration could represent $1 trillion of the global GDP, and $300 billion of the annual U.S. GDP, accounting for 1.6 million jobs in that country. Part of the excitement lies in the wide range of potential applications from business management to smart materials, and the accessibility of the idea: a nature-inspired concept is often engaging and could offer new solutions to old problems.

As the flexibility of the topic is extensive, the relationship between education and bioinspiration is also growing, especially when we make the most of its ramifications, differentiating between the technical (biomimetics) and the social/environmental (biomimicry) aspects. Institutions all over the world are introducing courses featuring these studies. For the Western part of the world, biomimicry has strong establishments of certified institutions such as the Biomimicry Institute.

---

6 Bagley, Rebecca O. 2014. "Five Tech Trends That Can Drive Company Success." Forbes. Forbes Magazine. April 1, 2014
Bagley, Rebecca O. 2014. "Biomimicry: How Nature Can Streamline Your Business For Innovation." Forbes. Forbes Magazine. April 15, 2014.
7 "Lessons of Design Learned from Nature." 2011. The Economist. The Economist Newspaper. November 9, 2011.
8 Murray, Sarah. 2018. "Biomimicry Unleashes Wave of Materials Innovation." Financial Times. Financial Times. January 8, 2018.

In developing countries and emerging markets, experts and institutions are moving fast to formally introduce these topics into curriculum as they provide innovative means to face challenges and also encourage the creation of multidisciplinary teams for collaborative research. According to FBEI, "bioinspiration's global spread represents a valuable dynamic in the global economy's progress over the next two decades as emerging markets continue to develop and expand at a faster pace than their industrial counterparts."[9]

Bioinspiration in Chinese Culture

If it has not been already demonstrated, sometimes it is not easy to draw a clear line between terms that share common language roots, this problem turns to be even more complex when trying to translate those terms to the Chinese language. One of the first challenges our lab took on was to not only deal with the confusion of defining the term within the field, but to also contextualize these meanings for a Chinese audience.

"*Biomimetics*" and "*Bionics*" have one expression that can be attributed to many areas. Its direct translation would be "all things similar to bionics", which in English could refer to biomimetic, bionics, biomimicry, bioinspiration, biological design, bioinspired design, biomorphism, biognosis, and biophysics.

In scientific papers, the word *bionic* is mostly used and it translates in Chinese as "仿生" (*fangsheng*). If we aim to break down its two characters, the first one (fang) means imitate or mimic, whereas the second (*sheng)* holds the meaning of life or biology. Therefore the combined implication from *fangsheng*, means to mimic life, synonymous with the English meaning. Similarly, if we look at the roots of the word bionic, it originates from the technical term bion (pronounced bee-on) (from Ancient Greek: βίος), meaning "unit of life" and the suffix -ic, meaning "like" or "in the manner of", hence "life-like ".[10]

9 Bioinspiration An Economic Progress Report. Report. Fermanian Business & Economic Institute, Point Loma Nazarene University.
Commissioned by: San Diego Zoo Global, November 2013.
10 Vincent, Julian F.V., "Biomimetics: Its Practice and Theory."

On the other hand, the word "biomimetics" can be also found translated as "仿生" in academic papers, but this is not widely used. And in the case of biomimicry, there is no actual translation that implies its environmental implications, this being one of the reasons why biomimicry is difficult to grasp by researchers in China. In general, the distinction between the different words appears to depend on the context and background of the research, rather than on a definition.

China, between Pollution and Innovation

Since the late 1980s the Asia Pacific region has surpassed both the United States and Europe as the main emitter of carbon dioxide into the atmosphere. The combination of fast growth and high population of countries like China, India, Indonesia, Pakistan, Bangladesh and Vietnam, has made this part of the world responsible for almost 50% of the global carbon dioxide emissions today, as shown in the BP Statistical Review of World Energy 2018.[11]

Interestingly enough, according to the Da Vinci Index, a database tracking scholarly activity in the field of bioinspiration[12], Asia is also the world leader in research publications related to nature-inspired design. In 2012, the United States and China accounted for 25% and 23% of the articles published in the field, respectively. Nevertheless, five of the top ten universities represented in bioinspired research papers were Chinese, against only two from the USA. In the same year, Asia accounted for 37% of the research, followed by Europe (29%) and USA (26%). After three years, Asia had increased its share up to 41%, followed by Europe with 33%, and USA with only 17%. These facts are an indicator on how developing countries like China and India are aiming to counter the consequences of their dramatic development with investment in research,

11 Rapier, Robert. "China Emits More Carbon Dioxide Than The U.S. and EU Combined." Forbes. July 01, 2018. Accessed December 06, 2018. https://www.forbes.com/sites/rrapier/2018/07/01/china-emits-more-carbon-dioxide-than-the-u-s-and-eu-combined/#7d437c29628c.

12 "The Da Vinci Index & Biomimicry." Point Loma Nazarene University, www.pointloma.edu/centers-institutes/fermanian-business-economic-institute/da-vinci-index-biomimicry.

innovation, and funding for programs which explore new sustainable methods from materials to infrastructures. In the case of China, in addition to these efforts, as far as spatial design disciplines such as architecture and urbanism are concerned, nature-inspired solutions are already an important part of the government's development program. The awareness of rapid urbanization being a main source for China's present environmental challenges has triggered nation-wide initiatives such as the generic eco-city program, and the more specific "Sponge City" program, which focuses on efficient water management and flood risk mitigation.

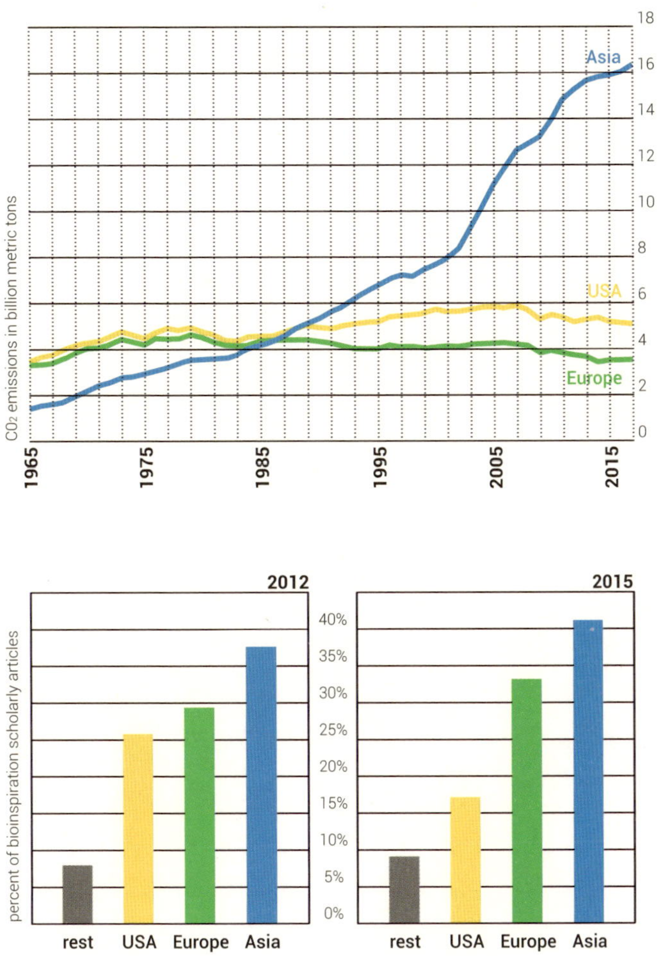

CO_2 & Bioinspiration

Top chart showing CO_2 emissions in billion metric tons from 1965 - 2015.[11]

Lower chart, percent of bioinspiration scholarly articles. in 2012 and 2015.[12]

Using again the 2013 report by FBEI, it is projected that by 2030, bioinspiration could help mitigate the potential economic costs of global resource depletion and pollution by about 9% or $500 billion.

The current bioinspired research being conducted in China varies per institution and ranges from the highly scientific work in nanotechnology to the conceptual design of artifacts through sustainable methods. This range has allowed for a broader discussion about bioinspiration within the context of this book, rather than solely focusing either the technical or the sustainable aspects of the term. The showcased work comes from laboratories founded by some of the top universities in China, namely the Key Laboratory of Bionic Engineering from Jilin University with work by Yan Liu; the Key Laboratory of Bioinspired Smart Interfacial Science and Technology from Beihang University in Beijing with work from Yongmei Zheng and Lei Jiang; the Dynamic Technology Lab for Bionic Materials and Structures from Donghua University led by Dengteng Ge; and the Biomimetic Design Lab (BiDL), from Tongji University in Shanghai led by Pius Leuba and Mary Polites.

Although it is true that most bio labs in China are currently focusing in biomimetics or bionics, we hope that the idea of innovation in sustainable development starts to get traction in both research and education. This collection represents only a small cross-section of the immense amount of research being produced by Chinese universities today, but it is a good sample of the current point of interest in bioinspired design in this country. In this regard, this book aims to continue to contribute to the development of nature-driven design in such a challenging but also inspiring context.

Lichen

Image Opposite Page:
Cladonia pyxidata,
"Pixie Cup Lichen"

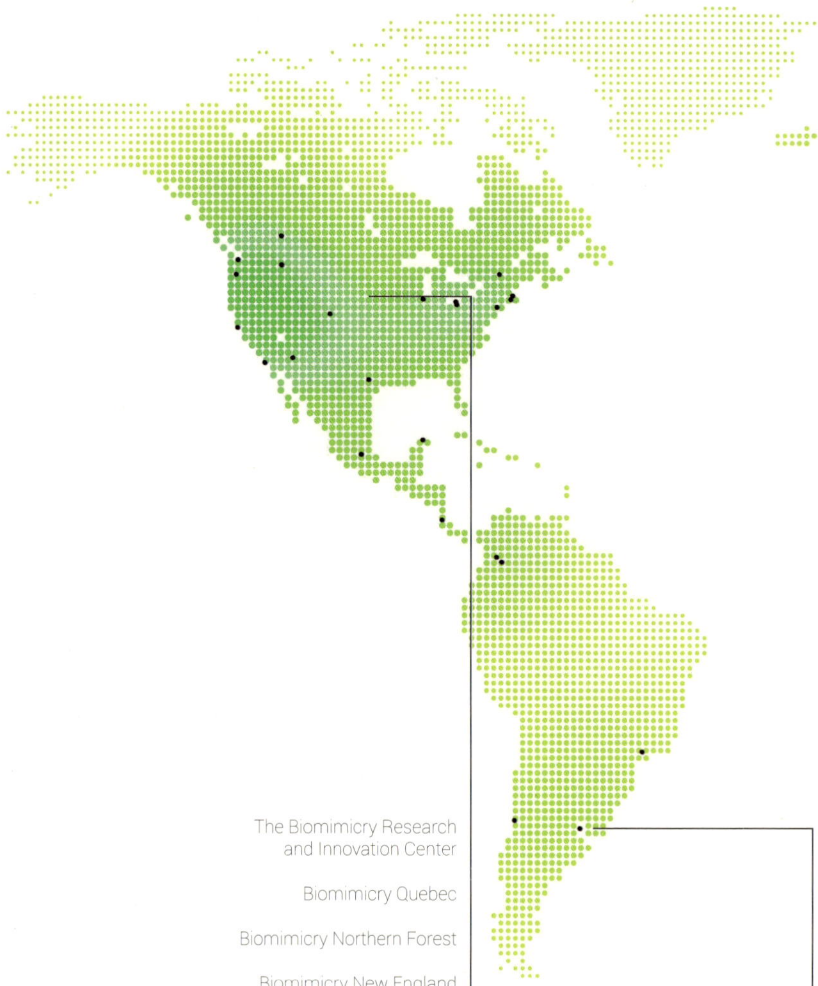

Biomimicry Labs Map

Concentration intensities of biomimicry related labs. Europe and the United States hold the most biomimicry labs concentrations. Interesting nodes in Central/ South America and Oceania. Map is for illustrative purposes only. For full list of labs, see appendix.

Lab Concentration

High *Low*

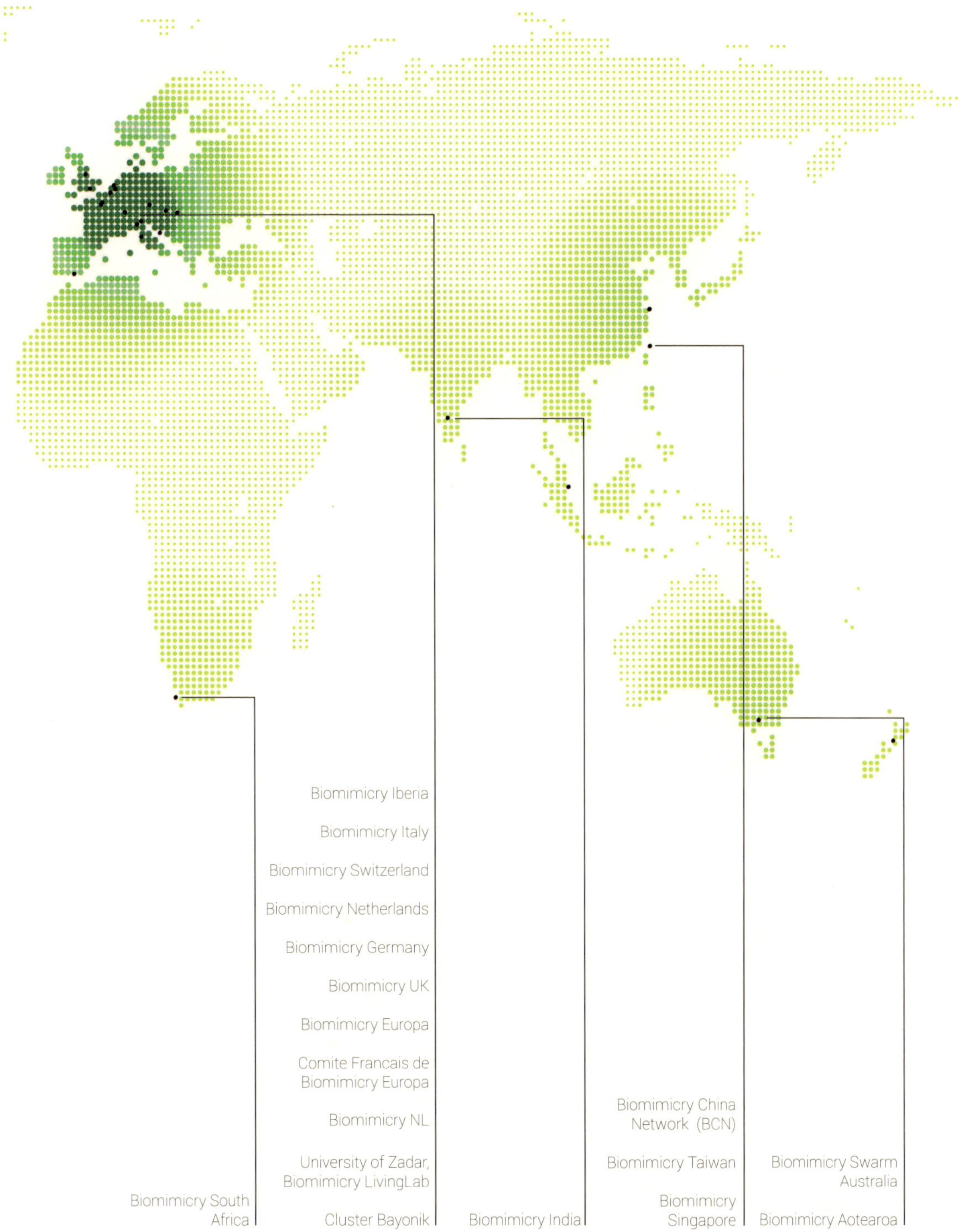
Biomimicry Iberia
Biomimicry Italy
Biomimicry Switzerland
Biomimicry Netherlands
Biomimicry Germany
Biomimicry UK
Biomimicry Europa
Comite Francais de Biomimicry Europa
Biomimicry NL
University of Zadar, Biomimicry LivingLab
Cluster Bayonik
Biomimicry South Africa
Biomimicry India
Biomimicry China Network (BCN)
Biomimicry Taiwan
Biomimicry Singapore
Biomimicry Swarm Australia
Biomimicry Aotearoa

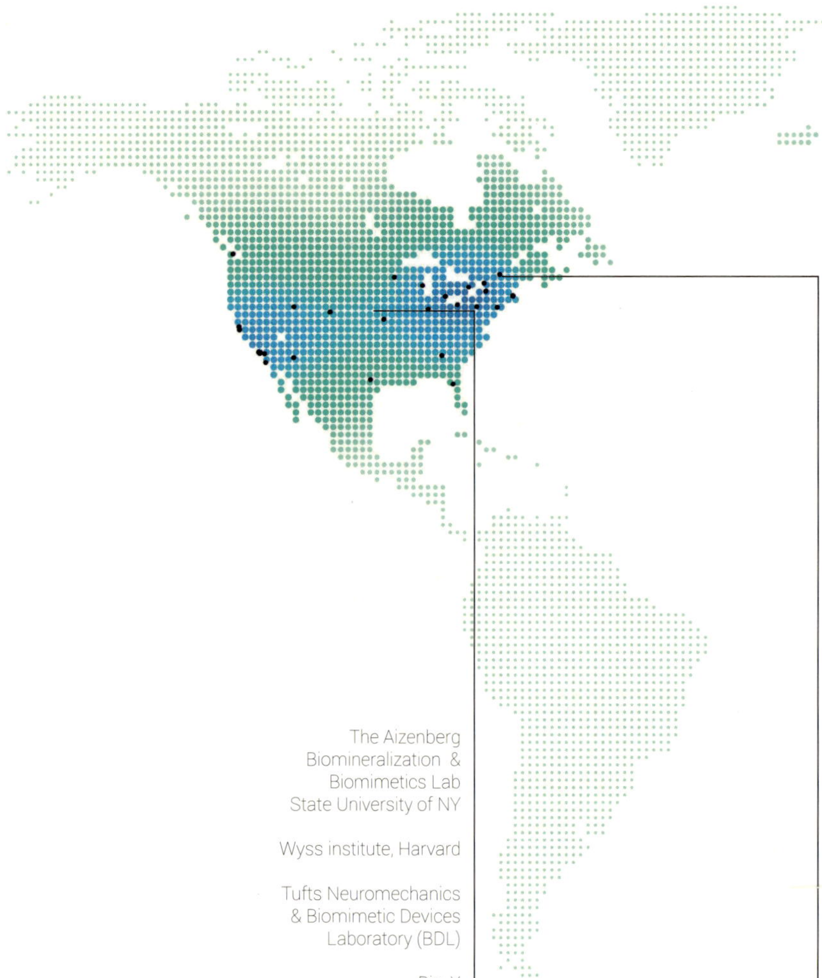

Biomimetics Labs Map

Concentration intensities of biomimetic related labs. As biomimetics was established in Europe, more labs are found there, especially in Germany. Still high concentrations in the USA, followed by China and India. Map is for illustrative purposes only. For full list of labs, see appendix.

Lab Concentration

High *Low*

The Aizenberg Biomineralization & Biomimetics Lab State University of NY

Wyss institute, Harvard

Tufts Neuromechanics & Biomimetic Devices Laboratory (BDL)

Bio-X

University of California at Berkeley, Bioengineering

The Biodesign Institute

Healy Laboratory

Biomimetic Research Laboratory Fergusonlab

The University of Utah Department of Bioengineering

Biomimetic Millisystem Lab

Mechanical Engineering Research Laboratory (Merl)

Ontario College of Arts & Design (Canada)

Queens University, Civil Engineering, Kingston (Canada)

Biomimetic Surface Engineering

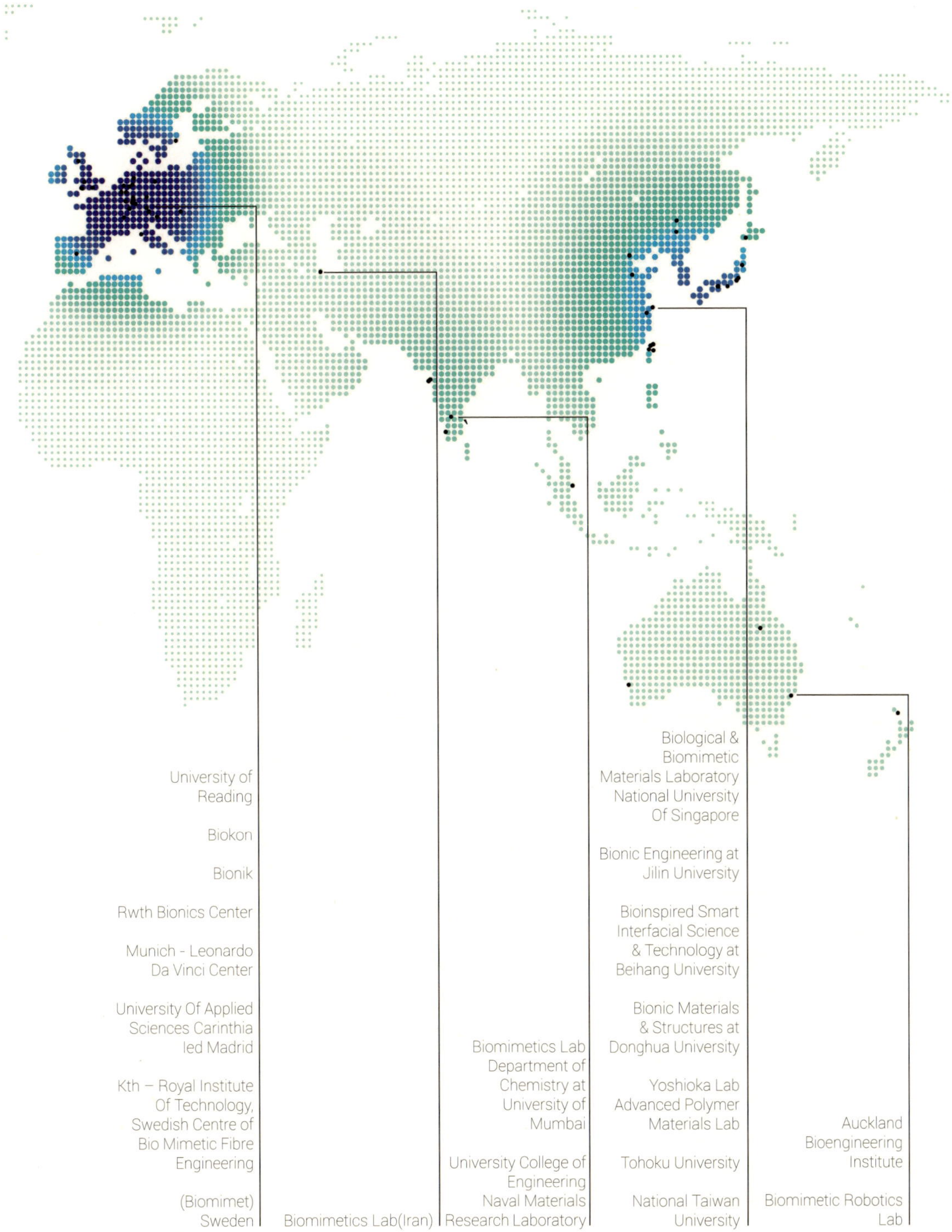
University of Reading
Biokon
Bionik
Rwth Bionics Center
Munich - Leonardo Da Vinci Center
University Of Applied Sciences Carinthia
Ied Madrid
Kth – Royal Institute Of Technology, Swedish Centre of Bio Mimetic Fibre Engineering
(Biomimet) Sweden
Biomimetics Lab(Iran)
Biomimetics Lab Department of Chemistry at University of Mumbai
University College of Engineering
Naval Materials Research Laboratory
Biological & Biomimetic Materials Laboratory National University Of Singapore
Bionic Engineering at Jilin University
Bioinspired Smart Interfacial Science & Technology at Beihang University
Bionic Materials & Structures at Donghua University
Yoshioka Lab
Advanced Polymer Materials Lab
Tohoku University
National Taiwan University
Auckland Bioengineering Institute
Biomimetic Robotics Lab

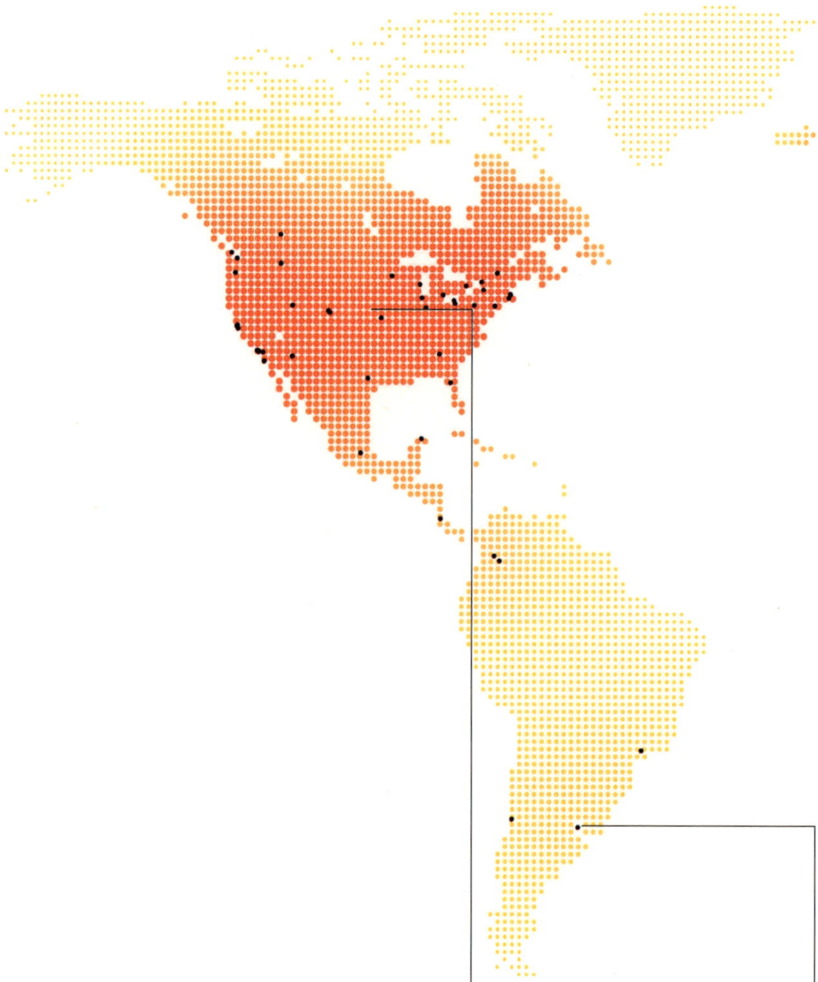

Bio Labs Map

Concentration intensities of biomimetic and biomimicry related labs. Bio labs are polarized between US, Europe and East Asia. It will be interesting to see the evolution of more isolated labs, namely those in South America and Africa. Map is for illustrative purposes only. For full list of labs, see appendix.

Lab Concentration

High *Low*

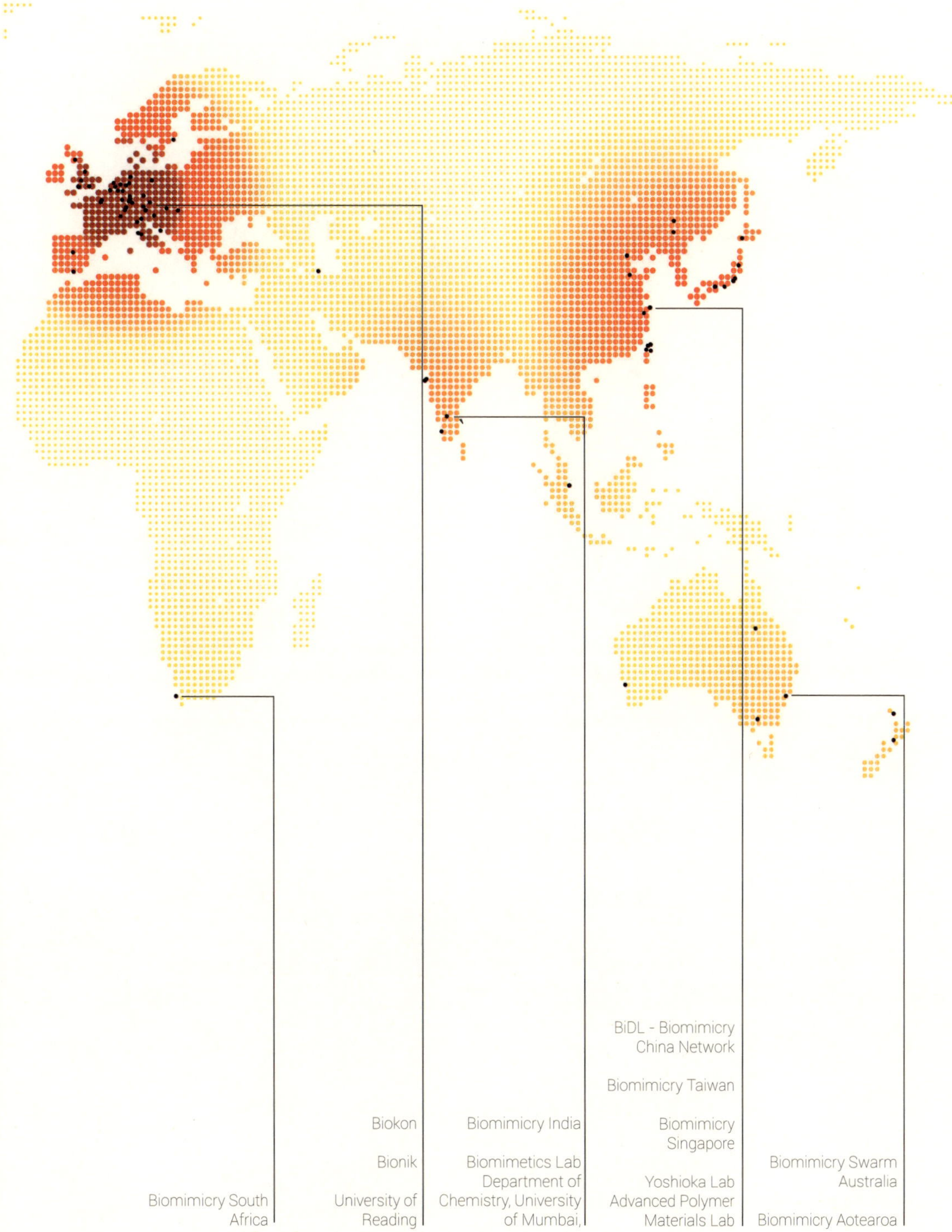
BiDL - Biomimicry China Network
Biomimicry Taiwan
Biokon
Biomimicry India
Biomimicry Singapore
Bionik
Biomimetics Lab Department of Chemistry, University of Mumbai,
Biomimicry Swarm Australia
Yoshioka Lab Advanced Polymer Materials Lab
Biomimicry South Africa
University of Reading
Biomimicry Aotearoa

Interviews

The line between the different nature-inspired fields is sometimes blurry. To explore the nuances within these approaches, the BiDL lab conducted a series of interviews with biological design experts with previous or current academic and professional experience in China. With a global background from Switzerland, United States, United Kingdom, and China, these experts' backgrounds range from natural or social sciences to design and engineering. We asked each interviewee to answer 10 questions covering their professional experiences and address what this means for current and future practices in China. The questions aim to showcase the range of implementations in which biomimicry and biomimetics are explored in education and professional practice.

Leaf Tendrils

Detail view of florets tendrils

Pius Leuba

Pius Leuba is an architect and founder of the biomimetic design lab (BiDL) at Tongji University. During this time he developed the regional Biomimicry China Network (BCN), which has been admitted into the Biomimicry Global Network. He contributed to the development of a culture assessment tool with BEE and GIGA that allows for the deduction of place-specific design strategies in locally adapted design. He currently teaches Climate Change for the Built Environment at the University of Liechtenstein.

Marie Dariel

Marie Dariel is the principal of Dariel Studio, an interior design company based in Shanghai and the UK. She founded IDEA Design Lab at Tongji University, a non-profit incubator promoting cross-disciplinary and forward-thinking approaches in China. This practice uses biomimicry and human wellness as drivers for change. She is the co-founder of the BCN with Pius and in partnership with the Biomimicry Institute.

Sherry Ritter

Sherry Ritter is a certified Biomimicry professional. After a career as a wildlife biologist, she started working with Biomimicry 3.8 in 2003 as part of the core team. Her research was seminal to the development of biomimicry in collaboration with businesses. She contributed to research which laid the foundation for the Biomimicry Institute and its educational materials which are used extensively.

Camille Fong

Cammille Fong is a Biomimicry Associate at Biomimicry Switzerland. Her work has been presented at the "Thoughts for Food" Summit in Zurich and at "Economy, People and Planet" in Copenhagen in 2016. She is currently studying international relations focused on the Cleantech sector and driving impactful collaborations throughout the globe.

Dengteng Ge

Dengteng Ge is a Professor at the Institute of Functional Materials at Donghua University, in China. His research is on biomimetic materials with dynamic properties from photonic ink to optical smart membranes derived from Cephalopods. The work of his lab has focused on functional materials and dynamic artificial microstructures for applications of wettability, smart windows, sensors and energy storage.

Yongmei Zheng

Yongmei Zheng is a professor at the School of Chemistry at Beihang University. Her research interests are focused on bioinspired surfaces with gradient micro- and nanostructures. She has published more than 90 SCI papers including significant journals such as *Nature*. She is a senior member of Chinese Composite Materials Society (CSCM), member of Chinese Chemistry Society (CCS), American Chemistry Society (ACS), Fellow member of NANOSMAT Society, International Society of Bionic Engineering (ISBE), and International Association of Advanced Materials (IAAM). She is a member of the editorial board of scientific reports in *Nature*.

George Jeronimidis

George Jeronimidis is a Professor (Emeritus) for the Composite Materials Engineering, School of Construction Management and Engineering at the University of Reading, UK. In 2009 he became Co-Director of the Emergent Technologies and Design program at the Architectural Association School of Architecture. George was elected president of the BIOKON International, in 2009. He became a fellow of the International Academy of Wood Science (IAWS) in 2005. He served as chairman from 2005-2013 as member of the Scientific Advisory Board of the Max Planck Institute on Colloids and Interfaces, Golm, Germany. He is currently on the Editorial Board of the new international journal Virtual and Physical Prototyping, Wood Science and Technology and Biomimetics & Bioinspiration.

Evan Greenberg

Evan Greenberg is an educator, designer and researcher. Since 2008-2017 he was appointed Studio Master of the Emergent Technologies and Design program at the Architectural Association School of Architecture. Evan is a Teaching Fellow at the Bartlett School of Architecture, UCL, where he led a postgraduate Unit 14 with Dirk Krolikowski. Evan has contributed to many international conferences and publications including the Association of Architectural Educators (2016), the Architectural Humanities Research Association (2015), the International Association for Shell and Spatial Structures (2015), eCAADe (2014) and Architectural Design (2013).

Julian F. Vincent

Julian F. Vincent is a zoologist. In 1968 he started his career at the University of Reading. In 2000 he was invited to the University of Bath, as a Professor of Mechanical Engineering. He was a part-time lecturer at the Royal College of Art & Design and Imperial College London until 2010. He is and has been, a member of numerous scientific and advisory boards. He co-founded the Center of Biomimetics in Reading and Bath and is president of the International Society for Bionic Engineering. He is an Adjunct Professor in Engineering and Materials at Clemson University, USA. He has published extensively on biomimetics including *Structural Biomaterials*, *Biomimetics: Its Practice and Theory*, and *Biomechanics-Materials: A Practical Approach*.

Pius Leuba

Could you give your own definition of Biomimicry?

To learn from nature, how humanity can ensure its endurance on planet earth.

How did you get involved with Biomimicry? When and how did you start learning about Biomimicry?

While continually exploring and developing the topic of sustainability since 2002, I felt somewhat sobered by the slow progress. Around 2010, I read the book *Biomimicry: Innovation Inspired by Nature* by Janine Benyus. Its inspiring and promising message captivated me like nothing ever before. But I couldn't readily figure out how to accurately apply this novel approach in my academic work and design for human sustainability in general. In 2012, during a week-long IDEA Design Workshop with Dariel Studio, we invited Ms Seema Anand, a certified biomimicry specialist, for a day-long input about Biomimicry. The biomimicry philosophy seemed the most convincing approach. I ultimately decided to establish the Biomimetic Design Laboratory (BiDL), at the Tongji University College of Design and Innovation.

Considering your background, which field(s) are you applying Biomimicry principles to (energy, structures, design, science, education)?

I am a trained architect, and my primary field of action is the built human environment. Anything regarding architecture, from planning processes to building and human spaces is my primary target in applying biomimicry. As a university professor, however, my top priority is to spread ideas of biomimicry. Once young professionals have tasted the potential of this mind-changing approach to human sustainability, they will, in turn, figure out a multitude of application fields. By inspiring others, I can have a more substantial impact overall than by just applying biomimicry in my own personal projects.

What are you working on or researching right now?

Besides the daily ongoing research projects with our students in the lab, I am interested in questions about the correlation between true sustainability and biomimicry. Nature is 100% ecological, but nevertheless rather cruel and pitiless. After all, it is a world of constant "eating and being eaten". From a humanitarian perspective (from which every individual has the right to life, fairness and happiness), biomimicry might primarily serve to make our societies and cultures more ecological, efficient and durable, but not necessarily more sustainable in a human civilization sense. Does true and complete human sustainability even exist? Can biomimicry lead us there? And if not, what is missing and what approach might offer a methodology to get us there?

What are your educational experiences with biomimicry?

As for my teaching, the first hurdle is always time. It takes a few hours or even days to transfer and understand the game-changing potential of biomimicry. Most people are usually somewhat astonished about the possibilities that unfold before them once they understand the philosophy, functioning and scope of this approach. Many people take a while, as it is a genuinely new way of thinking – truly outside the box. The second hurdle is the actual application of biomimicry in our daily work.

Due to the novel way of thinking and working, we are very clueless, slow and clumsy in the beginning. We don't know how to pose questions towards nature or how to get information and answers from it. We suddenly realize that we must start cooperating with diverse other professional fields, most of all biologists, just to design something as simple as a pencil. Implementation of biomimicry is something that many people take weeks and months to achieve – luckily it is much more fun, exciting and rewarding than the current routine way of working.

In lectures and workshops (hours to days), I try to primarily reveal the inspiration, potential and hopefulness of biomimicry. If there isn't much time to study the application, then the primary goal must be to convince attendees of the usefulness and value of biomimicry. Interesting examples help to achieve that. If they get inspired and enlightened, they will find their own ways of applying it. In courses and semester modules (days to weeks) however, I teach biomimicry as much as learning as by doing. I often keep the end-product or even the end-goal open-ended. Once attendees begin understanding the full potential of biomimicry, they start having their own application ideas. My role is then rather slow down and guide their enthusiasm and teach them methodical applications as well as conducting authentic research and implementation processes. Often the first outcomes are not necessarily satisfying yet, as in the process must also realize what a long way we have ahead of us. However, my experience is that most people are persuaded of the approach and want to keep working in that direction.

Where do you think people/students have the most success concerning learning and applying biomimetic principles? Where do they have the most issues?

I don't think I have seen a typical area so far, where people succeeded more than in others. It always depends on which life question, design problem, or natural process or artifact catches a person's fancy and triggers a momentum in exploring and learning. I used to think that people would be most successful applying biomimicry in their own field, but this requires a different way of thinking, and some people have trouble departing from their usual routine. This path is possible but not the only one. Quite the contrary, biomimicry might lend itself to produce solutions in unknown fields, where one isn't prone to fall into old ways of thought. It is said that innovation occurs at the junction in different fields of existing knowledge. Biomimicry requires, by definition, a few different professionals to join any project. This creates an ideal breeding ground to mutually pollinate ideas.

Design Week

Tongji University Design Week 2014 Biomimicry Workshop introduction lecture led by Pius Leuba

Having said this, I think the more natural way to understand biomimicry is through its philosophy, i.e. the way of thinking, outlook on life and design. Grasping its concept may occur as an astonishing revelation and expose huge potential and inspiration, which makes this step more rewarding and pleasant. The implementation of biomimicry, on the other hand, isn't something one can actually learn. All that the biomimicry methodology (Design Lens) offers is a way of approaching projects, but not ready solutions. One must start from scratch again with each new project. On second thought, this is evident: since every artifact has unique requirements, conditions and context, each one is by definition unique. To be completely sustainable, an outcome must consider all specific aspects of its place, time, resources and life cycle – just like nature does. Every approach to a new project is a new road tread upon and can't be learned beforehand. This is possibly what makes biomimicry the most holistically sustainable approach of humanity to date.

Considering biomimicry within the context of Asia and its current challenges (rapid urbanization, increasing population, pollution), where do you see these principles most likely to be explored?

Asia's problems are admittedly enormous. However, they are only a more intense manifestation of wrongdoing that has existed in the developed world for decades at smaller but growing scales. Namely, China, which has been learning (copying) a lot from the West since 1978, is merely illustrating what happens when you accelerate and densify western ways of living. Our so-called developed human culture is exceptionally unsustainable and unethical.

Where do you think biomimicry will have the most significant impact in the future?

Our most essential sensors, our eyes, are heavily limited regarding discerning things smaller than a certain size. Therefore, up to now, we have primarily studied and designed artifacts at a rather large scale from a natural point of view. Nature, however, has solved most of its problems at a much smaller scale than we try to address our own obstacles. We are just now in the process of starting to design at the micro and nanoscale and rendering things ever smaller (like microprocessors). I believe that this design realm of the small scales will reveal a world of new solutions and design outcomes for generations to come. Solutions might allow us to leave the path of "heat, beat and treat".

However, these might often be materialistic and mundane outcomes. Where I really hope that biomimicry is going to have an impact is firstly on our forms of governing ourselves, in such directions as open, fluid and fairer democracy; secondly, on our forms of work and exchange, in such directions as shared economy, equitable work and income, as well as unconditional basic income for every human being; and thirdly, and most importantly, on new ways of learning, such as equal access to shared information, open source learning and critical thinking.

I believe that biomimicry offers ways to not only avoid our harmful lifestyles but to regenerate and develop Earth's natural environment and its vital systemic relationships and processes. Asia must turn away from appropriating Western principles and remember its own values and sources of innovation. Together we could change the impact of human life and development. We could be the benefactor species of planet Earth.

Marie Dariel

Could you give your own definition of biomimicry?

Biomimicry is the imitation of the models, systems, principles and elements of nature to solve complex human challenges. It is about mimicking nature to create and innovate. Which means to reconnect with nature and life to find inspiration and rethink strategies, to make them more adaptive and efficient, correctly as nature has done for the last 3.8 billion years.

How did you get involved with biomimicry? When and how did you start learning about biomimicry?

My first involvement with Biomimicry was in 2006 when I launched a forward-looking lab for the engineers of Eiffage (France's third-biggest construction group) to develop innovative and sustainable concepts for urban systems in 2030. I organized weekly meetings with life scientists, and one of them came to talk about biomimicry. At the time, most people were smiling at the concept of mimicking nature to develop engineering solutions.

Then, when I moved to China in 2012, I created the same type of cross-disciplinary platform for young Chinese architects and designers to collaborate with life scientists and develop innovative designs with professional designers: it was called IDEA Design Lab. During one of the international workshops we organized at Tongji University, we invited one expert of biomimetic architecture from India, Seema Anand, and one architect-engineer from ARUP, Francesca Galleazzi who talked about cities like forests. Based on these talks, we asked participants to produce biomimetic solutions and designs inspired by nature for a futuristic urban hotel. We met great enthusiasm from the Shanghai community. By the end of 2012, together with Prof. Pius Leuba and Tania Conforti, we developed the first network entirely dedicated to Biomimicry in China and with the support of the Biomimicry Institute in the US, we started to work on the first workshop on Biomimicry ever organized in China, in a remote biosphere reserve of UNESCO. This movement began to grow, attracting interest from professional designers, but also from other universities eager to collaborate on this emerging science. The three of us gave several lectures and talks on biomimicry in Shanghai. This is how the creation of the BiDL developed and run which was then supported within Tongji University's College of Design and Innovation.

Considering your background, which field(s) are you applying biomimicry principles to (energy, structures, design, science, education)?

"We started to work on the first workshop on Biomimicry ever organized in China, in a remote UNESCO's biosphere reserve"

Until my move to the United Kingdom in 2013, I had always used biomimicry as a technical and scientific approach applicable to my fields: urban design, urban systems and architecture, although, my background being really pan-disciplinary, I have always been very much interested in all the other applications in engineering, energy, education, and the social sciences.

In the UK, I have studied for two years of training in psychology, more particularly in Gestalt psychotherapy, and many times I have found it interesting to apply biomimicry principles in this field. For example, the Gestalt approach to change is very close to the natural processes of resilience, flow and growth. Ultimately, understanding nature may bring a better understanding of the self and interactions with others.

What are your educational experiences with biomimicry?

I gave several lectures on biomimicry in China between 2012 and 2014 at Tongji University and Shanghai Normal University, as well as in Senegal, Burkina Faso and Kenya within the framework of the Cities and Climate Change Initiative, developed by UN HABITAT. I also organized several cross-disciplinary and hands-on workshops for professionals in Shanghai during which participants came from diverse backgrounds. They were divided into multi-disciplinary sub-groups and asked to develop innovative concepts based on biomimetic models previously presented by several speakers and biomimicry experts.

How do you approach teaching/speaking/introducing to others on biomimicry?

I think biomimicry can be presented to any audience, be it academic or professional, technical, scientific or philosophical, young or even younger. Observing nature is something anybody can do. But this observation needs to be carefully facilitated to raise awareness on processes and principles and then start building knowledge, bridges, and explore applications. I like to use simple examples from nature to convey an idea: the ecosystem of trees, for instance, the genius of the cactus' structure optimizing shaded areas to cool down, the self-healing miracle of skin or plants, the cycle of life. The demonstration needs to start from simple examples everybody knows but then to focus the attention on unknown chemical processes or mechanisms to reveal nature's engineering genius. It always brings smiles of amazement on people's faces. That is how you know you have inspired them, or not.

Where do you think people/students have the most success regarding learning and applying biomimetic principles? Where do they have the most issue?

Generally, people don't have difficulties playing with examples in nature and in every workshop I've facilitated, I always have mixed participants' backgrounds on purpose to allow for more debate and dialogue, people always brought up proposals. Nature's processes are so diverse that they allow people to be creative and the degree of technical mimicry can be very loose. I guess the difficulty would be then in the development phase as it may require rigorous work and constant collaboration with people who are not used to making things differently.

Considering biomimicry within the context of Asia and its current challenges (rapid urbanization, increasing population, pollution), where do you see these principles most likely to be explored?

I think Asia has a tremendous competitive advantage in its history and philosophies: for centuries, they have nurtured a better connection to nature, a closer relationship to its rhythms and rules, a stronger awareness to its principles and strategies, even though China's 20th century's history and recent economic development have taken its people away from this traditional culture and mindset.

But China needs to reconnect with its cultural DNA. Chinese policymakers know they have to slow down its hyper-consumption path, its unplanned urbanization and that they must develop qualitative approaches to sustain an increased and wealthier population. Therefore, I see biomimicry as a wonderful opportunity for them, as well as for researchers, engineers, designers and leaders, in general, to support and develop solutions that are more efficient, and in particular: consumption habits that are more reasonable (copy nature to produce less waste, or not at all, to ultimately upcycle everything, to achieve carbon neutrality, and so on). I get the sense there are proactive civil societies generally in Japan, South Korea and China is eager to follow nature's genius to innovate.

Sherry Ritter

Could you give your own definition of biomimicry?

Biomimicry is learning how to improve human designs by learning from nature. In my "elevator speech", I always jump right into examples, like the lotus plant inspiring roof tiles that self-clean when it rains and spider webs that incorporate UV-reflectant silk strands that have been emulated in glass for commercial building windows to reduce bird collisions.

How did you get involved with biomimicry? When and how did you start learning about biomimicry?

I've always been fascinated by plant and animal adaptations, that is, what strategies they have for surviving in their environments. I have a Master of Science in Wildlife Ecology and worked for many years for state wildlife management agencies. I often gave talks to school children and took them on field trips, usually focusing on adaptations. When I moved to Montana, I met Janine Benyus and heard about her book. In 2003 when I was less than a full-time employee, I asked her if she had any work I could do on a part-time basis, and she and Dayna Baumeister made an arrangement for me to collect biomimetic case studies. A few years later, I was hired as part of the Biomimicry Guild (now Biomimicry 3.8). I also worked for the Biomimicry Institute.

Considering your background, which field(s) are you applying biomimicry principles to (energy, structures, design, science, education)?

I've done a lot of projects with for-profit and non-profit sides of Biomimicry 3.8. I should say that I don't have any design background. Whether I was doing consulting, working on AskNature, or teaching workshops, my primary focus was on the natural side—helping clients and others learn from nature and how to apply what they learn to their designs. I also taught biomimicry methodology in workshops.

Throughout your work, were any of the projects realized?

One of the issues we always had was that we would work with clients, but we didn't always know what they did with the information. They might have come up with something, but we often didn't know what they developed. Recently Biomimicry 3.8 put out information that revealed some of the products that came out of consulting with them, however I'm not certain on which tasks I helped that directly translated into these products.

The largest report that was made public was the one we did for HOK which can be found online as the *Genius of Biome* (the link which can be found in the resources index). It seems that HOK did take some of those ideas further; I know they went into designs. Tom McKeag of HOK, who came over to the first public lecture in Shanghai at the beginning of the time I was there at the D&I and he gave some examples of things they were developing. For example, we talked about beaver dams—one of the strategies in the Genius of Biome—that hold water on the landscape, which helps to reduce floods but also helps provide ecosystem services like replenishing groundwater and keeping water cool. The design strategy that came from the biological strategy came down to using a series of leaky structures to slow down water. Tom suggested a concept of a series of rooftop structures, and even install things on the sides of buildings. These structures would reduce the amount of runoff going into the city water systems and decrease the number of pollutants reaching rivers, lakes, and the ocean.

Another example comes from before my time, but when Dayna and Janine were working with the Interface, and the result was the entropy carpet tile that emulated the organized chaos on a forest floor.

To help designers and for use in workshops and our professional courses, we developed the *Biomimicry Resource Handbook* (the link which can be found in the resources index). We did a lot of on line instructional materials for the biomimicry design challenges for the Biomimicry Institute, which was mostly put together by Megan Schuknecht, Gretchen Hooker and myself. The whole idea for the design challenge was not only to generate great designs around these significant issues but also help students learn the process of doing biomimicry while they did it. Not everybody who wanted to do the challenge was located where someone taught biomimicry. The idea was for them to get up to speed on their own.

What are you working on or researching right now?

I retired in 2013, then continued doing contract work for two years which again involved doing research for client projects, teaching workshops, working on AskNature, and helping with biomimicry design challenges. Currently, I'm enjoying retirement and getting involved in local riparian restoration projects, conducting bird surveys, and leading student field trips to learn about wildlife.

What are your educational experiences with biomimicry?

I'm a Certified Biomimicry Professional, which I earned mainly through work and teaching experience with the Biomimicry Guild and Institute. I've taught biomimicry design methodology in workshops in Mexico on three occasions, Costa Rica, China, and France, and taught two cohorts of the Biomimicry Specialist program, which were 9-month long courses taught on line and in-person sessions.

Can you expand upon what you taught in these international settings, how you dealt with language, concepts and give a general introduction?

The workshops in Mexico mostly included Mexicans, but also some Costa Ricans, Puerto Ricans, Colombians, Canadians, and Americans. The difficulty in Mexico for me was to some extent the language, as I wasn't well-versed in Spanish. We taught in English, and then when the students got into smaller teams, they would talk in Spanish unless there was a student from the States in their group. I tried to stay out of group conversations unless I felt a need to jump in and guide them, but with everyone chatting in Spanish,

I didn't know if they needed my guidance. It became an informal rule in a way, that when Sherry came close to the group, they needed to switch to English so I could see where they were and if there were any problems I needed to address. The other instructors were fluent in Spanish, and there was a lot of support from both directions, understanding what I was saying as well as what others wanted to say.

How do you approach teaching/speaking/introducing biomimicry to others?

Because I'm so in awe of nature's adaptations for survival, that's the approach I tend to take to start working with students. I want designers to think outside of their own human-centred knowledge and realize that rather than look to other human designs and ideas for inspiration, they can widen their scope to include possibilities they never dreamed of but that nature has "designed" through millions of years of evolution. This sometimes means that people struggle for a while; they're not sure why they have to learn about biology. But eventually the genius of nature's survival strategies becomes more apparent to them, and they too become inspired. I love the moment when they suddenly go from perplexed to curious and excited.

As an example, during one of our workshops we were working with a team from an oil company. For one of the tasks we needed them to verbalize what it was they really wanted to do, to think deeply about what their challenges were, and then state it as a verb. This task is exactly what the Biomimicry Taxonomy on AskNature is about, thinking about a challenge from a functional viewpoint. They had these functions in their minds e.g., protection from excess water, manage impact, so we took them outside and asked them to look at how nature does the same thing. They came back to the room so excited to share what they had found, and from then on they progressed quickly, looking at their challenges and potential solutions in a new direction.

I also think it's important to teach Life's Principles (from Biomimicry 3.8) or Nature's Unifying Patterns (from the

Biomimicry Institute), which are nature's ubiquitous strategies for long-term survival. Whether or not a client's or student's initial interest is in sustainable design, it's important for them to realize that nature's strategies are in the long-term and that by emulating those strategies, the client or student will not only create more sustainable designs but also find inspiration that can lead to better designs.

Where do you think people/students have the most success regarding learning and applying biomimetic principles? Where do they have the most issue?

"That's what the Biomimicry Taxonomy on AskNature is about, thinking about a challenge from a functional viewpoint."

I find that the ability to go of what we know is difficult. Sometimes our minds get in our way when we're trying to do something new. Young children get biomimicry right away. College-age students are a little more inhibited, but thinking about biomimicry comes easily to them. Older groups struggle more because they've been taught and practice approaches to design handed down through past generations. But when they open their minds to new experiences, they also often come around to this new approach to design. That transformation is a real joy to see.

One of the activities we often start with when working with students and clients is to get them outside and ask them to sit still for 20-30 minutes, studying one animal, plant, fungus, whatever catches their sight. They get restless after a few minutes, figuring they've learned everything and want to move on. We encourage them to stay and soon they "let go of what they know" and really start a period of deep observation. I remember my first experience doing this. I'm a biologist, so I already know a lot about nature. But after I watched a fly on a plant for as long as 20 minutes, I started noticing patterns of veins in its wings, how the structures

In-Field Exploration

Previous Page: To prepare students to understand the functions of nature without preconceived ideas – students are led in teams to explore nature and see how they understand functions through sensory activities

on its eyes and wings created different colors, and more details. It was a transformative experience, and every reluctant person I've sat down to do that activity has come away transformed in how they look at nature.

One of the things we did in France, and I do it with every group, is to try to get them to observe nature to the point where they get past the fear of not knowing, as it bothers people what they don't know. I would tell them that not understanding exactly how something works doesn't matter at this stage of the process. Many times we're out in nature with bad internet reception. For example, in Mexico, we didn't have access for eight days. The students have to be able to make observations and figure it out for themselves. As a teacher, I try to make them feel comfortable, and I do have a lot of scientific papers on my laptop, so that helps. I remember one time in France we went out for an exercise where I tried to get them to think about function. One student couldn't handle guessing and trying to figure it out for himself. He wanted to look it up in papers. But I was trying to get across that for this activity you don't need to know what is going on with a biological strategy. For example, think about what the function is of a plant's with fuzziness. Then you think about how you could apply function that to design on a basic level. It's really about getting them to feel comfortable with their observations; it's another one of those transformative moments, realizing that one doesn't have to know; they can discern and try to figure it out for themselves. Then later if they're going to go into design in depth, that's when they need to read a paper or talk to a biologist.

Regarding scientific papers, I wrote an article for the design challenge sections on line about how to read a scientific paper (https://toolbox.biomimicry.org/references/read-scientific-paper/). If understanding these types of papers isn't possible, then it is best to bring in a biologist. However, to work with a biologist or a naturalist you need to go with a particular question: "How does nature ___?". You need to know what kind of questions to ask. If you come to them with a scientific question and explain biomimetics to them, biologists get it right away.

One of our students from the biomimicry specialist course, Carlos Rego from Portugal, was a materials designer. Before he became a biomimicry specialist, he worked with his company to design a water bottle based on how white bark pine trees have a twist that gives extra strength in the wind and against heavy snow. He did that before he took the class and the design that came out was a lighter weight bottle that was very strong, but that used less material. But it was still plastic. Then he went on trying to incorporate life's principles into the company's design process so that they could rank each design as to how well it met those principles. Moving on to the next project, he dug into scientific papers and books to understand the material compositions. However this example is one where the transition worked. But, many times there are those that can read technical writing, and then there are others that are better off consulting biologists.

This transformation I mentioned is a success story. But going beyond designing things, I love it when I follow up with former students online or at meetings and discover that they are thinking much more deeply, looking beyond mimicking structure to mimicking processes and taking a systems-level approach to biomimicry.

One issue we've had is getting students away from bio utilization and bioassistance, which are different from biomimicry. In biomimicry, we're trying to learn from the living thing, to find out what makes it so strong, so lightweight, so resilient. We already know how to use the plants and animals (bio utilization and bioassistance). We use wood to build our homes, harvest animals and plants to eat, and use silkworms to create fabric. Biomimicry is going beyond what we know, expanding our minds by emulating, not using, nature. There's nothing wrong with bio utilization and bioassistance when done in a sustainable manner. But when teaching biomimicry, we want them to set bio utilization and bioassistance aside so we can go much deeper into understanding nature's strategies.

Sometimes we have a hard time getting students in our workshops to move away from engineering a problem the old way. It's what they're comfortable with, and we spend a lot of time in workshops pulling them back from that direction and returning to asking nature how to do it.

Considering biomimicry within the context of Asia and its current challenges (rapid urbanization, increasing population, pollution), where do you see these principles most likely to be explored?

Language itself is a barrier. Biomimicry and the wording and jargon we use in the West may not communicate well with English-speakers in Asia, and may not translate well into other languages because of cultural differences. I say that for two reasons. One, a group from Japan became interested in Biomimicry, but they had no word that worked for the term "biomimicry". Second, the wording and jargon created by biologists hasn't translated well to engineers, architects, and others, so I suspect they don't translate well into other languages and cultures.

Keep in mind that except maybe for India where I lived for two years in the late 1960s, I have very little experience in Asia and with Asian culture, so I can't speak with any authority, but here are a few thoughts. What I noticed when I spent a few weeks working with the BiDL in Shanghai and on our trip to Tianmu Mountain Nature Reserve is that Chinese people love exploring nature—a good start for learning in the natural environment. But in Shanghai and other large cities where many academics, architects, engineers, and other designers live, there is very little connection to nature that isn't in a cultivated form (i.e., gardens, arboretums). At Tianmu Mountain, workshop participants loved the way we explored nature and were so open to learning. While out on a hike, I sat on a rock outcrop looking over a forested hillside. One of the students watched me for a while, then came over and asked me what I saw while I looked out there. He wanted to see the view through my eyes, so I explained that I was looking at patterns in the landscape and pointed them out.

I've been disappointed that when we worked with international clients, especially in Asia, we didn't consult with their local biologists. These are the people who know their own countries and ecosystems the best and should be involved in projects. This is a matter of knowledge, but also of respect for local knowledge of both nature and culture.

Regarding China, I think to have success incorporating biomimicry you need to have some early success stories, from small projects to large ones. You could start small with a few helpful, valuable projects that can be shown to people. Then you could look at some of the huge problems and find solutions. Even if the solution is expensive, it would be worthwhile if it saves people's lives or increases their standard of living. For example, you could take a systems approach to address significant problems like flooding.

Where do you think biomimicry will have the most significant impact in the future?

Chemistry. The way we do chemistry may be the most extensive and most devastating disconnect between how nature makes things and the way we do. Nature's chemistry is life-friendly, while ours is often the opposite. We're enamored by our cleverness in manipulating chemistry, and too often heedless of its effects. We can learn from nature on how to do chemistry, such as how abalone makes ceramic stronger than anything we can make using proteins and minerals found in seawater without high pressure or temperatures. We can learn how to grow crops without poisoning the soil and killing microbes.

Resilient cities. I love the efforts to make cities function like ecosystems. I was talking to a few people from Beijing about their efforts to improve the air quality inside their office buildings so that the pollutants from outside don't affect people's health. I suggested that maybe they should also be thinking of ways those same buildings' exteriors could be designed to clean up the air. Recently, I've seen some articles talking about designers trying to do that. Even in these conversations, just knowing about systemic ways of thinking starts to address broader issues—these are all steps forward.

Camille Fong

Could you give your own definition of biomimicry?

Biomimicry is an approach that combines art and science to rediscover nature's beauty and intelligence to find solutions to challenges in our society. By looking at nature which has over 3.8 billions of years of evolution or "research and development", biomimicry can help us engage in innovation. The definition of biomimicry itself goes beyond the concept of sustainability because nature is a form of life which can evolve and regenerate.

How did you get involved with biomimicry? When and how did you start learning about biomimicry?

The first time I heard about biomimicry was in 2013 during a sustainable workshop organized by Moana Lebel, founder of Institut du Biomimétisme francophone in Montréal, Québec. A few weeks later, I met Arndt Pechstein, co-founder of Biomimicry Germany where I start learning about biomimicry. His passion for biomimicry was so contagious that he became my mentor and inspiration to integrate biomimicry into my projects. The best way to learn about biomimicry is to apply it. So the same year, I decided to participate in the Global Biomimicry Student Design Challenge organized by the Biomimicry Institute 3.8. This design challenge allowed my team and me to dive deeper into biomimicry methodology and find a life-friendly transportation solution.

Considering your background, which field(s) are you applying biomimicry principles to (energy, structures, design, science, education)?

With an environmental engineering background, I seek to find a holistic approach to today's societal challenges by integrating biomimicry thinking into the engineering problem-solving mindset and methods. Engineering is moving slowly towards a more integrated approach which includes sustainability. However, there is still much to be done in our understanding of sustainability and by bringing nature-inspired solutions I believe engineering will have a more significant impact. I have mostly explored biomimicry principles at the academic level.

What are your educational experiences with biomimicry?

Besides applying the technical aspect of biomimicry in engineering, I have also given hands-on workshops and talks to a general public audience on biomimicry to spark their interest in this innovative field.

How do you approach teaching/speaking/introducing to others on biomimicry?

When I introduce biomimicry to a general audience, my aim is that people will think differently about nature after my talk. I make sure that the audience knows that biomimicry is not a new concept, in fact, it was used a long time ago. Leonardo da Vinci applied biomimicry to study birds in the hope of enabling human flight. But back then, it wasn't framed as biomimicry. I believe that this is an essential statement because by knowing what biomimicry is, some of us may have used it in the past and with those lessons learned we could collect more information and contribute to biomimicry knowledge. I think it is also important to mention that we are not the only intelligent species on this planet, as there are more than 9 million species and we should remain humble towards them as they survive and thrive on earth millions of years before humankind. They are a book of knowledge, and we should learn from them before it all disappears. I believe that by seeing nature as a teacher, we will respect them more and get inspired.

Where do you think people/students have the most success concerning learning and applying biomimetic principles? Where do they have the most issues?

Biomimicry thinking is very innovative because it brings back students to ways of discovery that sparks their curiosity about nature. The first step is to bring them to a natural environment and observe the surroundings. This then leads to a questioning phase where students ask why this shape/color/form/pattern is the way it is. Since learning goes beyond a conventional academic setting, students tend to be more interested in the topic and understand better biomimicry because they use their senses including sight, hearing, taste, smell and touch to find reasoning. However, when it comes to applying biomimicry principles, a biologist in the room is often needed to truly understand the science behind it and avoid preconceived ideas. Also, working in the biomimicry field requires work in a multi-disciplinary team which may have some challenges as people work differently.

Considering biomimicry within the context of Asia and its current challenges (rapid urbanization, increasing population, pollution), where do you see these principles most likely to be explored?

My understanding of the context of Asia is still at an early stage, but I think that China is looking into more efficient innovative solutions which can be scaled up within a few years. Chinese investors want to see results and growing numbers and are not willing to take risks for research and development which have not yet proven its business strength in the Chinese market. I think that progress and experiences from the West on biomimicry is crucial to implementing biomimicry principles in Chinese industries since China tends to get inspired and/or replicate successful business cases and adapt to their context to fuel their economy. Applying biomimicry principles is a long process and in China time is money, as everything moves at a fast pace. With optimism, I believe that once accepted in the market and understood by intellectuals, the biomimicry approach will inspire many like-minded people on how to maintain a balance of energy and matter flow in a rapidly changing environment.

Where do you think biomimicry will have the most meaningful impact in the future?

I would say that biomimicry would have the most vital impact in the future in helping us maintaining an equilibrium between resilience and efficiency. This is how we're going to adapt to this new changing environment and thrive on Earth. By focusing on needs rather than economic growth, biomimicry could provide guidance and different alternatives to solve societal problems.

Lecture

Camille Fong speaking on Biomimicry

Dengteng Ge

Could you give your own definition of biomimicry?

For something to be considered biomimetic, it should inherit the benefits from natures' advancements, primarily material mechanics, structures or functional mechanisms. A key point, is to learn about natures' mechanisms and generate new ideas, but not to completely copy as any artificial applications is by default distinctly different in its environment.

How did you get involved with biomimicry? When and how did you start learning about biomimicry?

I have a background in materials science and engineering, so the jump into natural systems was easily connected. In the beginning of my career it happened that we compared synthetic materials with natural one as a basis of comparison. It was in my master's degree that I started to learn about biomimicry through reading scientific literature.

Considering your background, which field(s) are you applying biomimicry principles to (energy, structures, design, science, education)?

In my lab, our research is looking at how to apply biomimicry principles to functional designs and fabricate materials and structures. Specifically work focusing on the reaction of light and pigmentation which can be see in the projects in the Natural Prototype chapter.

What are you working on or researching right now?

Our research is about functional material and dynamic structures, which is for optical and mechanical applications. The latest work we completed was exploring dynamic structures which focus on movement can change a material, can be induced do to movement, similar to functions found in Cephalopods, or Octopus.

Considering biomimicry within the context of Asia and its current challenges (rapid urbanization, increasing population, pollution), where do you see these principles most likely to be explored?

I believe that transportation and urban design may benefit the most from the implementation of biomimicry principles as they are issues where the country is seeing the most demand due to the developments in China.

Where do you think biomimicry will have the biggest impact in the future?

In my field (materials science), many researchers explore the same fundamental concepts such as the hydrophobic surfaces of the lotus leaf or the hydrophilic properties of a Gecko's foot and the structural color of the Blue Morpho butterfly, for a few examples. Although though these are popular concepts – new finding help all of us to develop our work in new ways. From this, I think that the future wouldn't necessarily have a specific area of big impact, but larger innovations across the entire field due to improvements from tech and material explorations.

Yongmei Zheng

Could you give your own definition of biomimicry?

I think that biomimicry can be understood as an inspiration for designs in a variety of ways such as materials, surfaces, structures, functions, and so on through studying natural creatures. For example, when we studied spider silk, we learned that it could be rebuilt as a surface to act as water collection. When this concept is paired with a form that has curvature, the spider silk can achieve water collection abilities. To explore this through prototyping, we designed artificial silk, similar to the surface of wet spider silk. From this process, this is what we think could be described as biomimicry.

How did you get involved with biomimicry? When and how did you start learning about the field?

My fascination came from an innate interest in nature throughout my life. I understood only in my studies the extent to which interesting natural phenomena are all around us, e.g., lotus leaves with self-cleaning properties, strong spider silk, and butterfly wings. Although these features fall within other disciplines of science, such as physics and chemistry, materials, math, and building models, they all are contributing towards the design of novel concepts. To pursue these as a scientist and researcher is to engage with all of these studies and is what interested me the most.

Considering your background, which fields are you applying biomimicry principles to (energy, structures, design, science, education)?

Based on my background, I most frequently apply the principles of biomimicry to the design of surface structures and their related functions, like water repellent surfaces, anti-icing, and fog-collecting/harvesting, which are presented in the Biomimetic Prototypes chapter.

What are you working on or researching right now?

My research interest, is learning how the methods of controlling wetness are applied in biology, which can have a larger range of applications than one might think. The inspiration for many of these projects come from the classic concepts. We have been prototyping artificial surfaces through various techniques that employ unique effects such as anisotropic structures (directionally dependent materials), how ultrahydrophobic surfaces can act as anti-icing, anti-frosting, water-repellent functions.

What are your educational experiences with biomimicry?

"My interest is in discovering the secrets of how biology uses methods of controlling wetness"

Biomimicry is an all-engaging topic which became part of my PhD research and then transitioned into my postdoctoral research. During this work, I developed techniques that allowed for us to observe the unique structures on biological surfaces through new methods. These findings provided our work with a platform that helped us establish a formal lab and a means of how we could work together through the new method. Today, the lab is comprised of PhDs and other postdoctoral professors, interested in exploring their work with this platform.

How do you approach teaching/speaking/introducing to others on biomimicry?

I introduce the concepts in my courses at the university and while developing work with PhD students using examples of studies from our lab. Depending on the audience some topics are presented which are easier to grasp upon introduction, like bioinspired spider silk for the design of materials. These topics have easy departure points which are related to practical applications of water repellency.

Where do you think people/students have the most success concerning learning and applying biomimetic principles? Where do they have the most issue?

Biomimicry focuses on many aspects that are today properties of a variety of various disciplines. Research in my group are typically focused on the functions of micro- and nanostructures of surface and materials.

As materials and surfaces are the primary focus of my group, we develop work that requires a lot of specificity in physics, chemistry and nanotechnology. To fully realize a project, our team must have the capability of speaking to these fields. For some, this is one of the most challenging aspect. Many researchers are highly specialized and biomimicry requires them to play many roles.

Considering biomimicry within the context of Asia and its current challenges (rapid urbanization, increasing population, pollution), where do you see these principles most likely to be explored?

Despite the advancements of bioinspired materials in China and across the world, the abilities of these advancements to become commonly accessible remains a challenge as science is always a bit ahead. As with other topics, such as rapid urbanization, increasing population, pollution, there could be advancements in development within the past five years or even today that could mitigate or begin to assist in addressing the pressures from these topics. However, as I can only speak to my field in terms of biomimicry, I would suppose that if biomimicry were to address these topics, it should be to satisfy requirements associated with materials. Which, although might sound like a trivial topic, the idea of pollution is surrounded by materials, waste and water, which play major roles in contributing to the global problem, as we all know.

"As materials and surfaces are the focus of my group, we develop work that requires a lot of specificity in physics, chemistry and nanotechnology. To fully realize a project, our team must have the capability of speaking to these fields. Many members of our team are researchers that are highly specialized in topics but biomimicry requires them to play many roles."

Yongmei Zheng PhD - Professor at the School of Chemistry, Beihang University

George Jeronimidis

Could you give your own definition of biomimicry?

To me, it's honestly no question; biomimetic design is an abstraction of good design taken from biology. There are number specializations at various levels, such as material, structural, fabrication, and a lot of other aspects, but it's about trying to find out how biology works and to understand what things we can use in our own way.

How did you get involved with biomimicry? When and how did you start learning about biomimicry?

My background was in mechanical engineering, and when I went to the University of Reading, I started working in composite materials. At that time, there was a colleague of mine who was interested in wood. He suggested that wood was an interesting biological composite and that I should have a look at it. That's how the whole thing started. For me, with interest in fiber systems and hierarchical biology systems, biology was a minefield of information. That's how I got involved. During the first few years of work, although we were working on biological materials, we were also trying to understand how they work, and in some cases applying the knowledge we had at the time. That was before biomimetics and biomimicry as it is known today existed as an established discipline, as this was around 1980s.

At the time there was nothing specific in biomimetic research; there were a few pockets of research and innovation but no community as such. At the University of Reading we were probably the first ones in Europe working on this sort of topic. At the same time there was a harboring interest in the states. Biomimetics as a well-known topic started after 1980 when the US government set up a large research program title bioinspired materials. This program helped to bring the community together; before then there were a few people scattered across the world. The first conference on biomimetics was in 1980, or

1981 perhaps, and it was quite useful because it allowed the community to get in touch and develop ideas and exchange views. From this point on, work related to this area acquired the "biomimetics" label.

Considering your background, which field(s) are you applying biomimicry principles to (energy, structures, design, science, education)?

"My interest has always been to look at materials which could respond to changes in the environment such as temperature and humidity"

The main interest of my work has been on materials, structures, systems and smart materials as well as smart structures. And by smart I mean material systems and structures able to respond to inputs and adapt, which are the key to success in biology. I have always looked at materials which could respond to changes in the environment such as temperature, humidity, pressure and other factors. When we began, this meant moving away from the passive technologies we had at the time and to try and inject more functionality using concepts and ideas derived from biology. From my experience in doing this type of work, I think that design should be done in a broad context which includes specialties like advanced materials design. Its important to not just consider the materials but the material as part of a more integrated design concept.

Throughout my career, teaching has played a crucial role as it is always tied to my research. However, to provide some perspective, the field of biomimetics evolved slowly because at the time I started there were no official courses. The biomimetic world was doing research, with some isolated bits of teaching embedded into other courses. It was only after around 1995 when successors began to appear, that then there was a more coherent approach to try to produce teaching aspects of education in biomimetics. This runs in parallel to some extent with the fact that there were many science journals which started accepting papers in biomimetics. Which at the time, was struggling to find a forum for dissemination, so it was kind of a no man's land.

My colleagues and I started the first journal with a biomimetic title after the US conference in biomimetics. As we had the opportunity of surveying the level of research activity in other parts of the world, we thought there should be a journal that would be a bit more dedicated to the subject, because otherwise, people would find it difficult to publish due to the complex nature of biomimetics. This first journal started and lasted for a few years, and then other journals, more prestigious than this one and with longer traditions started adding sections on biomimetics. There are now at least half a dozen journals, some of which are published in China, and a couple in the UK. I've been more involved in teaching and research, but just as active in publishing journals and informing the science community.

How did you connect your research to design side and outcome? Was it a reaction to materiality aspects and where there was potential?

In a sense, it was a response. After you work on a material the next question is what you do with it. We could see the amount of materials people were inventing and developing and then afterwards addressing the question of what to do with it, which was not very effective. For us, it was to look at the material as part of the entire system similar to the biological world. It's almost impossible in a biological process to put a clear demarcation between what is material,what is structure, and where to place the boundary, as it could be anywhere. Which is not the case with the traditional approach we know which makes steel is steel, concrete is concrete, glass is glass, and we put it together to make something. Biology goes another way.

What are you working on or researching right now?

As I'm partially retired, I'm not involved in any active research, but I'm still involved in teaching. One of my primary interests is in architecture, and on concepts of smart technologies, smart materials and responsive materials. Specifically in the last few years, my main focus has been on bio-inspired robotics and on sensing and sensory perceptions. These aspects are fundamental if we want to design responsive structures as the senses are part of the system. You can't respond if you don't sense. The complication begins when you aim to integrate these two functions in the same way as biology.

What are your educational experiences with biomimicry?

I have been requested to teach courses around the world, and many of this teaching came in the form of working with PhD students. Therefore it wasn't in such a formal setting, but rather through a lot of the teaching that I was able to improve my background through the projects that the students were carrying on themselves. These were relatively small groups focusing on specific topics, and even today the formal teaching of biomimetics is fairly rare. The first course that I'm aware of, which mentioned biomimetics in the syllabus, was at the University of Bremen in Germany. The course was started by a colleague who I worked with a lot at conferences and scientific meetings. She was an engineer by trade, and she was teaching in an engineering school, and she decided to have a labeled biomimetic component to the course. I'm sure there was a lot of teaching, but that was the first one specifically indicated. I think the core issue with this not being widespread as a topic at institutions is that sometimes it's challenging to justify new courses given the number of students who might enroll in such a course. This course I'm mentioning was early on when the community was very small. Nowadays, it has grown enormously, more in the UK and other parts of Europe. There is still not a lot of biomimetics formally taught in a sort of pre-packaged format. However, there is a lot of biomimetic teaching within existing courses as additional topics. The areas that are most commonly associated with biomimetics are people working in materials, robotics, and smart systems. They are all conducting teaching education research, but not indicated as biomimetics.

How do you approach teaching/speaking/introducing to others on biomimicry?

If you take the kind of communication in the subject, it might be through courses at the university, undergraduate, postgraduate levels or at conferences, as there are a lot of conferences on biomimetics at the moment, what is very interesting for me is the fact that there are a lot of students who take part. This type of education is not in the formal way of spreading information, but it is quite effective. For me, the way to do it is through examples. You need to show first off where biomimetics has been successful, however, few the examples may be, it is in these instances that you can abstract the necessary thought process. It is a different approach from a more traditional discipline. Every time groups have started working in biomimetics but not interacting with the biology it's been a disaster, the biological side must be kept alive. All the information is in the biology. This also means that to teach biomimetics properly, you need a combination of people with biological backgrounds. I'm not a biologist, what I have learned I have learned over 30 years working with biologists. Without them, it wouldn't be the same. I think it's important to emphasize aspects of biology as we are just scratching the surface, and even after 40 years, there are still so many things that we don't understand how they work. The connection with biology is one aspect, as well to work through examples and use the examples to show what indicate what should be the approach. The process is to look at and understand the biology, identify very clearly what is it you want to be inspired from and then look at your technology to how to develop it.

When we started work many years ago, we had dreams and thought wouldn't it be a nice "if", and after more than three decades these dreams have become a reality. One dream was simply being able to measure things on a much

Desk Review

George Jeronimidis at IAAC for review

smaller scale. The improvements in microscopy, atomic force microscopes, X-ray diffraction methods, all this stuff has made great strides and provides information that previously wasn't available. These developments oblige everyone to revisit ideas in the light of new inventions.

On the other hand, one fantastic development is the development of rapid prototyping. When we start talking about biological material systems, they are incredibly complex concerning their assembly of different bits, fibers, orientation, such that even when we could understand what all that was about, we couldn't replicate it. We couldn't replicate necessarily at the scale in which it had to be replicated. Whereas with modern fabrication techniques, especially in 3D printing, which has made enormous strides due to its scalability, groups are working with very small objects to be printed. These are the sort of things we were dreaming of 30 years ago. The improvement in the range of materials which can now be used to expand the potential of rapid prototyping. Twenty years ago, 3D printing was only plastics because the technology was only rudimentary and plastics were the only materials that could stand the tech. Nowadays we have metals, plastics, ceramics, and groups working on fiber laying techniques; micro-manipulation of fibers such that the abilities of 3D printing is one of the keys.

The point that is important for people to understand when you look at a biological system superficially is that it all appears very complex, because the architecture is complex, the systems, the chemistry, and there are so many inputs on the genetic side. However, what is essential is to illustrate is that the physical principles which operate are not very different. The complexity comes in the way biology uses these physical principles, not the physical principles themselves as there are no new physics in biology. Everything operates similarly to basic garden physics in many ways, but the way in which biology can use these physical principles and extract benefits from is where the complexity is, and this is where the challenge lies.

Where do you think people/students have the most success regarding learning and applying biomimetic principles? Where do they have issues?

I think students tend to be very excited in general. I remember years ago, I was asked to give a series of lectures on biomimetics to a school of management and I asked: *Are you sure you've got the right person?* That was a challenge because it became a question of how do you project your views in the minds of people who are interested in management. But then interestingly enough I remember there were several students, despite the very few of them had a background in the sciences, who were amazed. They were so responsive, so really the important thing for me is to show that it is a different way of thinking and people respond to challenges like that.

Considering biomimicry within the context of Asia and its current challenges (rapid urbanization, increasing population, pollution), where do you see these principles most likely to be explored?

In the last three or four years when I was asked to give a conference to groups in France and Italy, what I found interesting was an issue that came up—*can biomimetics do something for sustainability and the environment?* This is an apparent trend now, as very often people would like to see biomimetics able to help with these topics and in particular for countries in the Far East, with concerns of pollution, population growth; achieving this requires a different way of thinking. We're not going to solve these problems through traditional ways of thinking, and I'm not suggesting biomimetics is the only answer, but it's a certainly a potential contributor. It's a question of saying if we want to accept sustainability, how do we translate that go into a series of scientific, technical engineering, architectural appropriations. I think in using that if we want a sustainable environment, what we mean by it, and how therefore biomimetics can help achieve benefits. This is one aspect which has been in the back of my mind for a few years now, and it's interesting that these things come back again.

At the end of the day, if we want a sustainable environment, we need to have systems in place which are more efficient in energy. Collecting energy is one; we are hopeless in our technical world at storing energy, leaving nuclear power aside. For example, in the UK, you've got a serious situation in which a number of wind turbines are invested in; they are all located up in the north where the windy conditions are perfect. When the wind is blowing too hard, and when the turbines are generating at their peak efficiency, they have to be shut down because they are generating too much energy and the grid can't pick it up. People are starting to wake up to this. The reason is that as we can't store the excess energy and can't store it because it's been ignored. If you look at biological systems, whether it's plants or animals, a lot of the tissues in animals are storage, storage of fat or whatever else, and in plants, you've got starch. When you reach the scenario where you've got more energy at the moment than you can deal with, you don't throw it away; you store it so you can use it later on when you don't have it. To me, it's one of the biggest challenges, as currently, it isn't acceptable. What I'm saying is that all the things that are desirable from a sustainable point of view are hitting barriers like this one and that somehow these need to be addressed. For energy storage, one of the most exciting aspects of biology is how animals manage their energy input and consumption apart from being more efficient at least in part that they can store excess energy when they don't need it, and then tap it when they do. This should be studied in greater detail.

I think that there are a few little pockets of interest that are beginning to emerge on this topic. What is also interesting is that in biology, apart from a few exceptions, nearly 99% of what biology makes is made or less at room temperature. A lot of our processes, cement, concrete, steel, even plastics, all these things need to be processed at high temperatures. For a while now there have been a few people that have been looking at bio-mineralisation because shells and shellfish and whatever can produce amazingly efficient structures, which all happens at room temperature. But what is the trick? How can these things make these materials at room temperature? This is a big challenge because if we can make things at room temperature, our energy consumption goes down, so there is less pollution, and in a variety of ways, there are indirect benefits. For me, if I had to restart again, these are perhaps where I would become more involved. One is focusing on energy storage and looking at biology to try and find ideas for energy storage while the other one is on aggregation processes that can be made at room temperature as well as in mostly very benign environments as opposed to chemical.

"For energy storage, one of the most exciting aspects of biology is how animals manage their energy input and consumption."

One of the other challenges which I don't have an answer to is the question of pollution. One solution is to try to avoid it, that would be the ideal solution but realistically speaking that's not going to happen for a number of economic and social reasons. If we can't deal with the problem at its source, then we need to find ways to deal with it once it's happened. I would like to see if anybody is thinking about how animals live in clean environments. Even defining what they consider a polluted climate as for us, it might be clean, but for them, it might be very highly polluted. How do they deal with pollution? Our lungs perform a double function of being able to separate oxygen from the blood system, but at the same time acts to stop pollution from getting into the bloodstream. I'm sure there are tricks, how one could try to understand from biology which one may have pollution control.

Where do you think biomimicry will have the most significant impact in the future?

If you look at fabrication technology, 3D printing is going to be one of the biggest players because of its scalability. The other field to benefit enormously, especially in the last few years, is robotics, which grabbed biomimetics regarding sensors and actuator materials systems. Soft robotics for example, as opposed to hard robotics, is booming at the moment because people are beginning to realize the limitations of 3D systems. Robotics is going to carry on and grow significantly due to areas where it can be applied, mostly because the application of robotics is sometimes more obvious to identify. If you take medicine, for example, robotics in medicine has had an enormous impact and is moving very quickly, because now we have technologies able to use this information in smart robotic systems.

The other is the area in which people have been working on for several years, which is collective behavior, in which we can take a research project in robotics and make things super intelligent. However, in biology, this relationship to intelligence doesn't work that way. Biology works in that rather than having one very smart sensor, it has 1,000 cheap and simple sensors, and this is a trend which the robotics community is beginning to understand. We're talking about collective behaviors, swarm behaviors that run straightforward rules. In studying bees and termites, we have not gotten the whole picture of how these things do what they do. We know the cause, and the effect but how the mechanisms deal with this level of behavior isn't apparent. I think this is an area that will receive more attention in the upcoming years.

Computation is another important tool not only for individual elements but also when looking at this collective behavior, and some level of computation is going to be necessary; but not necessarily complex. That's the challenge, to get the computational side without all the heavy elements. Rolf Pfeiffer, at the University of Zurich, Zurich, is probably the first one that I could communicate with because he understood the fact that in biological systems, the emphasis of the performance is not on the computational side. In other words, the interest is in as little computation as possible and a lot delegated down the to the materials system. He was very clear that if you have the wrong materials, then you need heavy computation to control the systems. But if you have better materials that are responsive and adaptive, then you take a load off the computational side and start thinking about a more practical approach for collective behavior.

If you were to start over and explore biological concepts now, where would you be most interested?

Well to me, one aspect is to go back to fabrication at room temperature. Complex fabrication at room temperature will have an enormous impact on the sustainable field, and at the same time, it will be able to provide sustainable systems that are needed to perform many functions. This is one of the areas I would revisit, and where a lot of work should be done.

"Biology works in a way that rather than having one very smart sensor, it has 1,000 cheap and simple sensors. "

If we have fabrication at room temperature, we don't have to get specific production locations. It's a way of redistributing, which is why the fabrication, or applications for fabrication processes, is not concentrated in specific locations with high inputs and high pollution levels. If you want to make steel, you can't do it in a diffused kind of way. This is what we inherited from the industrial revolution, which to a large extent is still with us. We need to revisit the more we can use energy on a local level and not shift it from A to B and transmitting the energy over meters.

What differentiates biology regarding materials is A- the room temperature processes which are possible; and B- the source of energy is diffused and not concentrated. We are poor in using diffused energy sources. This makes energy capture, or a new way of energy capture, one of the topics I would research. Of course, we have solar panels and wind power, but these things are having problems as well because they are too energy-intensive or they hit barriers with storage of energy. You can collect the energy and then figure out what to do with it, either to use it immediately or waste it.

Then, considering energy harvesting and I'm leaving nuclear power out, as it's the only one completely different. If you think about where our energy is coming from, it mainly comes from what the sun has been pouring down over earth for billions of years. At the end of the day, there is no way out, and the only ones that can convert solar energy into fuel are plants. No plants, no energy. I find it difficult when people have problems with that because it's so obvious. Plants collect the sunlight, convert it into materials, and the materials which plants made using solar energy were converted into oil and fossil fuels, of which we're now exploiting. We don't have to go through that same cycle. In other words, it's more interesting how plants can capture the energy, which would allow for shortcuts in the system. Then we can go from solar energy to fuels without having to go through the geological time scales of producing coal or oil. To me, this is something that needs more thinking. I know there are a few photovoltaics going this way, but it's not quite the same, as they are designed to produce electricity from capturing energy from the sun, and then we're back to the same problem to whether using it at the time of production or losing it. We need to start thinking of biologically-inspired systems which can capture sunlight and rethink about what to do with the trapped energy.

In addition to my first points, in plants, these things also happen at more or less room temperature. This is one area if, given the opportunity, I would want to get involved.

"The only ones that can convert solar energy into fuel are plants.

No plants, no energy.

Plants collect the sunlight, convert it into materials, and the materials which plants made using solar energy are converted into oil and fossil fuels, of which we're now exploiting. We don't have to go through that same cycle. In other words, it's more interesting how plants can capture energy, which would allow for shortcuts in the system. We can go from solar energy to fuels without having to go through the geological time scales of producing coal or oil."

George Jeronimidis - Professor (Emeritus) at the Composite Materials Engineering School of Construction Management and Engineering at the University of Reading

Evan Greenberg

Could you give your own definition of biomimicry?

For me, biomimicry is using nature as a model in its scientific definition, as a way to understand a process—and not as a metaphor. Rather than trying to create forms, it's much more about abstracting logics, which could be related to geometry, mathematics, hierarchy, and materials. The most important aspect is how we can leverage that information for design. In an expansive sense, this could be applied to architectural design, engineering, product design and any areas where systems need designing.

How did you get involved with biomimicry? When and how did you start learning about biomimicry?

I was first introduced to peripheral ideas like emergence and self-organization in an abstract or conceptual sense by Jason Johnson while I was an undergraduate student at the University of Virginia. That wasn't biomimicry per se, but more of ideas from biology thought about in a design context. I became more interested when I studied at the Emergent Technologies and Design (EmTech) program in 2007, where George Jeronimidis (one of the pioneers in the field) was a lecturer. I think that opened up my mind to the possibilities of using biology not just for inspiration but as a model for design. It was no longer about concepts and discussing ideas; we can begin to think about these things and apply them in a real, material sense. As I have been working with George at EmTech, I would say I'm still learning from him every day.

Considering your background, which field(s) are you applying biomimicry principles to (energy, structures, design, science, education)?

In a very general sense, I have always been interested in biomimicry as a model for structures and material applications, broadly defined as integrated material systems. In my research, I have been thinking more architecturally, where performance and form are synthesized, and that has stemmed from eight years of active and passive response explorations in material systems. These responsive systems are part of what I have focused on in my teaching on at the AA for a long time. In the studio, we recently started to look at how biology could be a model for organizational systems, and that has led to computational design outputs such as agent-based modeling or branching algorithms, for example. This is with an attempt to broaden the investigation from the mechanical side, which the program has done with a high success rate over the years. In looking at computational systems, we are addressing more collectives or social organisms because the power of computation is that you can embed many complex relationships, which wouldn't be possible otherwise. With this approach, this means that with a lot more data, you will have more "moving parts" in a way when looking at things like ants, bees, termites, and flocking and other patterns of social behaviors. We had students looking at jellyfish for example. Even though the organism or organization might be complex, it's about abstraction. In my opinion, it doesn't matter how complex the organism is, but how good you are at abstracting the information.

What are you working on or researching right now?

In my current work, I'm looking at developing a spatial model for thermal comfort. This model aims to address where architects can design parameters over which they have control. While I don't think it's directly related to biomimicry, it comes from a broader interest in the way you find hierarchy and differentiation in biology and how this allows for robust systems.

What are your educational experiences with biomimicry?

I have been teaching in EmTech formally since 2009 but began assisting George in the Natural Systems and Biomimetics course since 2008. I think it's important to say that biomimicry isn't siloed in the seminar but is extended into all studio topics on a daily basis.

How do you approach teaching/speaking/introducing to others on biomimicry?

I deal with designers generally and architects more specifically, but across my academic years in the past I've taught the first year, fourth and fifth year, and post-professional students. I approach any student the same way. I'm a designer, and I'm teaching designers; our roles are not to be biologists; I say this because when I started, I felt like I needed to be one, and I think we're just kidding ourselves if we believe that we can be biologists as it is not what we are set up to do as designers. Architecture is different. I think it's important that people understand that we can learn from biology but that we are not trying to create it—we're not trying to design lobsters or pine cones. There are many who think that they will look at a turtle shell and think, "now let me make a turtle shell" (which is perfectly fine if that's the goal)! This is what I mean about abstraction. That's the most important thing for students and anyone to understand that we have to know enough so that we can make design decisions. Another important consideration is that biomimicry is as much a design tool as a pencil or script. In the end, it's just a tool, and we use it because we can leverage it in a way that allows us to think and execute it in entirely new ways, which is why it's useful.

In learning biomimicry, I think there are two significant struggles, one of which may sound bizarre. One major issue is that students feel as though they can't understand biology because science papers are too technical, and that becomes a roadblock. Understanding technical writing is the first thing to get over, and sometimes it's difficult because we are not scientists. But once overcome, and I think this is the bizarre part: students very quickly become intensely interested in and excited by biology. There is so much information that it can easily sidetrack us from the point of the exercise. We have to remember the thing that we are after is a design solution, so we can read papers for months and years and spend our lifetimes doing that, but abstraction is crucial. The most successful student projects are those where students can look at the biology, diagram it, draw it, model it, understand it, then throw it away.

Where do you think people/students have the most success concerning learning and applying biomimetic principles? Where do they have the most issue?

If a biological model can help you think about a problem, then you're just as well off as the person who can run a simulation or any other tool. I believe success in learning comes from having a clear goal and abstracting information as quickly as possible. A lot of progress has come with projects dealing with responsive material systems, due to a more direct application from biology to design. We are seeing a lot of exciting results arising from the use of evolutionary computation, which I think comes from increased accessibility to tools built for this purpose.

Considering biomimicry within the context of Asia and its current challenges (rapid urbanization, increasing population, pollution), where do you see these principles most likely to be explored?

In EmTech we are addressing issues present in China like population density and rapid growth, and if we return to be the adaptive idea and resilient systems that exist in biology, for me, I think that's where we can begin to look.

Computationally, we are capable of introducing large sets of data, energy flows and similar sorts of information, which can provide potential insights to address these specific issues of context. In this sense, it is more on a strategic level where there is room for a biomimetic design intervention, rather than a particular system application.

Where do you think biomimicry will have the most significant impact in the future?

I'm particularly interested in how biomimicry can help with spatial organizational logic. Considering this will hopefully help us deal with things like climate change or issues which will require rethinking our initial approach. I think this will require us to look at biology more for its strategic models, which need intensive studies to address our current systemic issues. If we can embed intelligence into building design, through organizational and strategic means, an integrated system design, we could potentially be better off.

Julian F. Vincent

Could you give your own definition of biomimicry?

When I was first asked this question, my off-the-cuff response was "The abstraction of good design from Nature". Now I realize that it's more like "The implementation of good design from Nature". But this needs broadening, in that you can't know which bits of "design" are going to be relevant to the solution of a problem (the same is true of any technology). I think there is further modification required which suggests that both the technical problem that you're trying to solve and your information about nature, are removed from their immediate context. The easiest way of marrying these two, I now realize, is defining problems, both those from technology and from nature, as trade-offs. However, you're still left with how to decide what the trade-off is in the natural system. Mathematically there are ways of defining this which we are starting to learn about (e.g. Pareto analysis and associated techniques).

How did you get involved with biomimicry? When and how did you start learning about biomimicry?

I prefer to call it biomimetics. "Biomimicry" in its current usage is somewhat biased by "AskNature". I am a biologist first and foremost who started out as an entomologist, and then insect physiologist then became interested in materials (initially insect cuticles). I branched out into other materials, primarily as a result of the community's rejection of my ideas and conclusions on the origin of the mechanical properties of cuticles. Those views are now endorsed, but it took 15 years, by which time I was a materials scientist working with engineers. This background in materials meant that the move from modeling materials theoretically to making physical models of the materials to test ideas.

Considering your background, which field(s) are you applying biomimicry principles to (energy, structures, design, science, education)?

All of the above. Nature doesn't recognize the limits that most people impose on it through different areas of knowledge. Neither should wc.

What are you working on or researching right now?

I'm developing an ontology for biomimetics which is included in my series of reports on ResearchGate and project titled, *Ontology of Biomimetics Based on Trade-offs*). This work is ramping up after about 6 years of concentrated effort!

My main interest has always been to try to make simple rules to bridge the gap between biology and engineering. This focus is what my ontology of biomimetics speaks to.

How do you approach teaching/speaking/introducing to others on biomimicry?

Case-based, it is the only way.

Where do you think people/students have the most success concerning learning and applying biomimetic principles? What are the principles?

There are none. Until there is a proper theoretical basis for biomimetics, there can be no principles which allow for the derivation of concepts and ideas. Biomimetics is largely about design. But then so is engineering where the math verifies the design. We currently have no such verification in biomimetics.

Regarding pedagogy, I have taught both engineers and biologists. The best biology students have no problems with math and can go on to be very successful with an unusual mix of skills. The more creative engineers can do very well, but biology is far more complex and, with no basic principles, is challenging to learn. Thus any part of biology that is not numerate, or at least can be stated in algebraic terms, will not be able to transfer to technology.

A case in point is the modern concepts of ecology. It is next to impossible to transfer ecology as natural history into a technical context. But we can express food chains as a matrix of interactions between predator, herbivore, and cannibal. There are mathematical methods available for analyzing matrices, so that some studies in ecology can be abstracted in a manner that is free of context, making it possible to expand the models into alien systems such as heat exchangers and factory towns. At its best the exchange of information is symmetrical, yielding novel models for both the biologist and the industrial designer. Unfortunately, most ecology is too complex for such a relatively simple model.

"My main thrust has always been to try to make simple rules to bridge the gap between biology and engineering."

Considering biomimicry within the context of Asia and its current challenges (rapid urbanization, increasing population, pollution), where do you see these principles most likely to be explored?

In China, I helped establish the International Society of Bionic Engineering (ISBE), based in Jilin University, and have been its president since its inception some eight years ago. I also helped establish the *Journal of Bionic Engineering*, now published by Elsevier. I have seen a steady increase in membership of the ISBE worldwide together with increased variety and quality of the research. However, I'm disappointed by the lack of biologists who take part—at only about 2% last I counted. The rest are engineers, physicists, mathematicians, or others. From the limited view, I have of China's developments, South Korea and Japan (the Asian countries where I have the most experience) I can't judge the impact of biomimetics on the social side. I know the interest is there, but the problems tend to be more complex and less constrained.

Lecture

J. F. Vincent lecturing at the University of Reading, Centre for Biomimetics

Where do you think biomimicry will have the most significant impact in the future?

There is a real chance that it will be rejected as useless. A few areas are embracing it within engineering—robotics (based on behavior), architecture (based on shape and function; a bit on materials). Biochemistry and medicine have been present in the area for a long time ("biomimicry" was first coined in the early 1960s as a concept in biochemistry; "biomimetics" was coined a little earlier in association with physics, robotics and computing). Most mechanical engineers regard it as a source of interesting but trivial toys. They find it difficult to cope with the complexity of biology, and with the lack of a theoretical and substantial basis. Engineering is about reliability; biology is about survival. These are very different philosophies. Sustainability doesn't need biomimicry since ecologists and soil scientists were there already; urban design might learn from it a bit, but I think we have yet to see the results. The problem with all of this is we see what we want to see, and that is often more influenced by social factors than by powers of observation.

"The best biology students have no problems with math (biology could not exist without statistics) and can go on to be very successful with an unusual mix of skills.

The more creative engineers can do very well, but biology is far more complex and, with no basic principles, is challenging to learn. Thus any part of biology that is not numerate, or at least can be stated in algebraic terms, will not be able to transfer to technology. "

Julian F. Vincent - Zoologist
Founder of Center of Biomimetics in Reading and Bath

Natural Artifacts

In biomimetics, there are several topics that researchers typically reference or build upon within their work. These include attributes, material properties, organization principles or shapes found in nature that recurrently inspire designers and scholars, eventually leading to applications in highly diverse fields. As a preface to the presentation of the labs' projects, this chapter introduces some of these concepts and links them to their most representative sources in nature.

Ribbed & Veined

Close up view of vein pattern on plant leaf

Hydrophobic
Surfaces that shed water due to micro scaled surface bumps

Hydrophobic & Hydrophilic

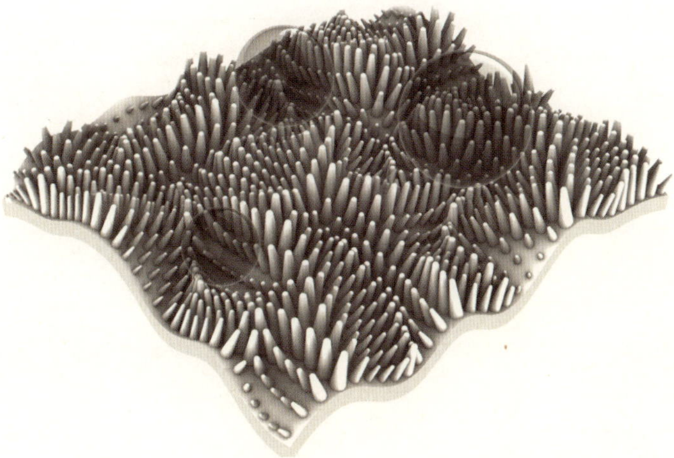

Hydrophobic:
Lacking an affinity for water; insoluble in water; repelling water.[1] Commonly found in plants surfaces used to assist in water directional transport and water shedding to clean surfaces.

Associated Terms: *Lotus Effect, Superhydrophobic, Ultrahydrophobic*

Hydrophilic:
Having an affinity for water; capable of interacting with water through hydrogen bonding.[2] Almost the opposite of hydrophobic. Commonly found in aquatic plants that are adapted to living in submerged conditions.

Associated Terms: *Oleophilic, Superhydrophilic, Ultrahydrophillic*

1 "Hydrophobic." Biology Online. Accessed November 22, 2018. https://www.biology-online.org/dictionary/Hydrophobic.
2 "Hydrophilic." Biology Online. Accessed November 22, 2018. https://www.biology-online.org/dictionary/Hydrophilic.

Special Surfaces

Opposite Page: Grass blades with water droplets depicting hydrophobic conditions.

Current Page : Diagram of a hydrophobic surface with water droplets

Phyllotaxis
Alternating patterns in many plants allows for growth to expand and solar exposure to be maximized

Phyllotaxis

Phyllotaxis:
The alternating pattern that is commonly found in plants allows for growth to expand as well as maximizing access to sunlight for petals and florets. The following description explains the mathematical relationships as quoted from "Wolfram MathWorld".

"The beautiful arrangement of leaves in some plants, called phyllotaxis, obeys a number of subtle mathematical relationships. For instance, the florets in the head of a sunflower form two oppositely directed spirals: 55 of them clockwise and 34 counterclockwise. Surprisingly, these numbers are consecutive Fibonacci numbers. Lilies, irises, and the trillium have three petals; columbines, buttercups, larkspur, and wild rose have five petals; delphiniums, bloodroot, and cosmos have eight petals; corn marigolds have 13 petals; asters have 21 petals; and daisies have 34, 55, or 89 petals—all Fibonacci numbers."[3]

Phyllotaxis Daisy Flower

Opposite Page: Florests of a Bellis perennis also known as Daisy, Bruisewort & Woundwort

Current Page: Diagram showing phyllotaxis pattern

3 "Phyllotaxis." From Wolfram MathWorld. Accessed November 22, 2018. http://mathworld.wolfram.com/Phyllotaxis.html.

Trichomes
Plant hair-like
structures

Hairs & Needle-Like Structures

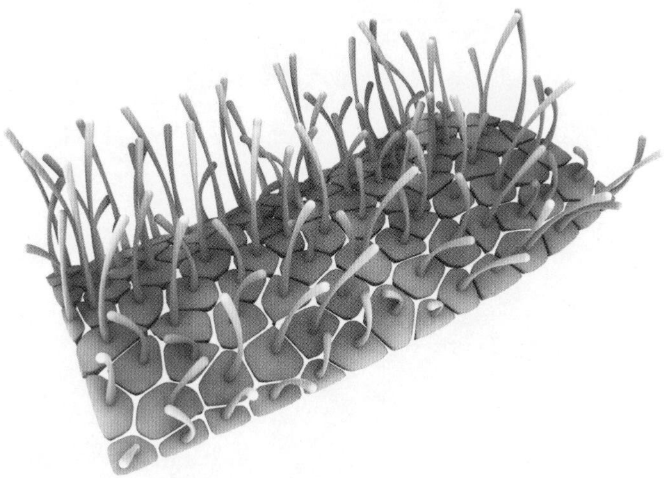

Plant and animal hairs and needle-like structures are often useful as their surfaces have unique specializations. The most widely known are the setae on the bottom of Gecko feet, which increase the Van der Waals force between its feet and surfaces. This force is what allows Geckos to walk upside down on surfaces.

Seta:

A slender usually rigid or bristly and springy organ or part of an animal or plant.[4]

Trichome:

(botany) An outgrowth (hair-like or scale-like) from the epidermal cell. (microbiology) A hair-like structure on the surface of certain microscopic organisms. More commonly trichomes relate to plants.[5]

Stem Trichomes

Opposite Page:
Image of strawberry stalk trichomes

Current Page:
Diagram showing conceptual trichome, setae, hairs and needle-like structures.

4 "Seta," Merriam-Webster, , accessed November 22, 2018, http://www.merriam-webster.com/dictionary/seta.
5 "Trichome," Biology Online, , accessed November 22, 2018, https://www.biology-online.org/dictionary/Trichome.

Nanofibrils
The cell walls of plants are built up with nanofibrils of cellulose

Micro & Nanofibrils

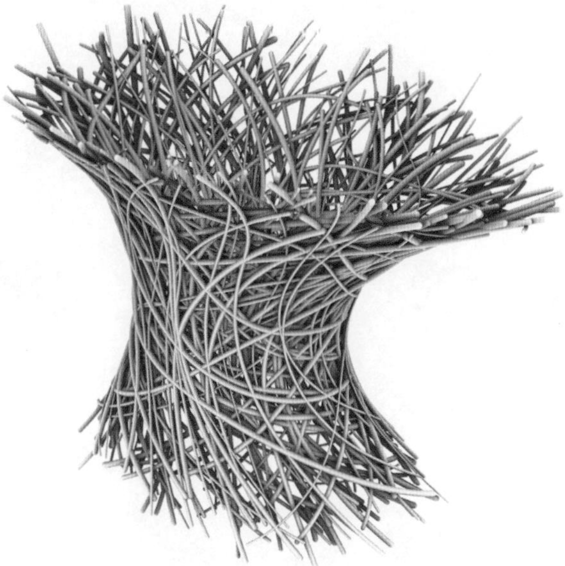

Fibers found in nature have provided insight into many fields from tissue engineering, adhesives, and engineered materials. Cellulose perhaps is one of the most common fibers referenced for its impressive hierarchical arrangements.

"Nano-scale refers to sizes between 0.001 and 0.1 μm (1 to 100 nm). This implies that the nanofibril term relates to fibrils with diameters less than 100 nm. Based on this definition, microfibrils can be considered nanofibrils, which also are composed of crystalline and amorphous regions. However, the difference between microfibrils and nanofibrils is that the former is a well-defined biological structure found in plant cell walls, whereas the latter can be considered a technological term introduced to describe secondary and engineered structures".[6]

6 Chinga-Carrasco, Gary. "Cellulose Fibres, Nanofibrils and Microfibrils: The Morphological Sequence of MFC Components from a Plant Physiology and Fibre Technology Point of View." *Nanoscale Research Letters* 6, no. 1 (2011): 417. doi:10.1186/1556-276x-6-417.

Cell Walls

Opposite Page:
Live leaf cells of the moss *Bryum capillare*, showing abundant chloroplasts (green spherical bodies) and their accumulated starch granules (elongated bodies within chloroplasts).

Current Page:
Conceptual diagram of bundled nanofibril arrangement.

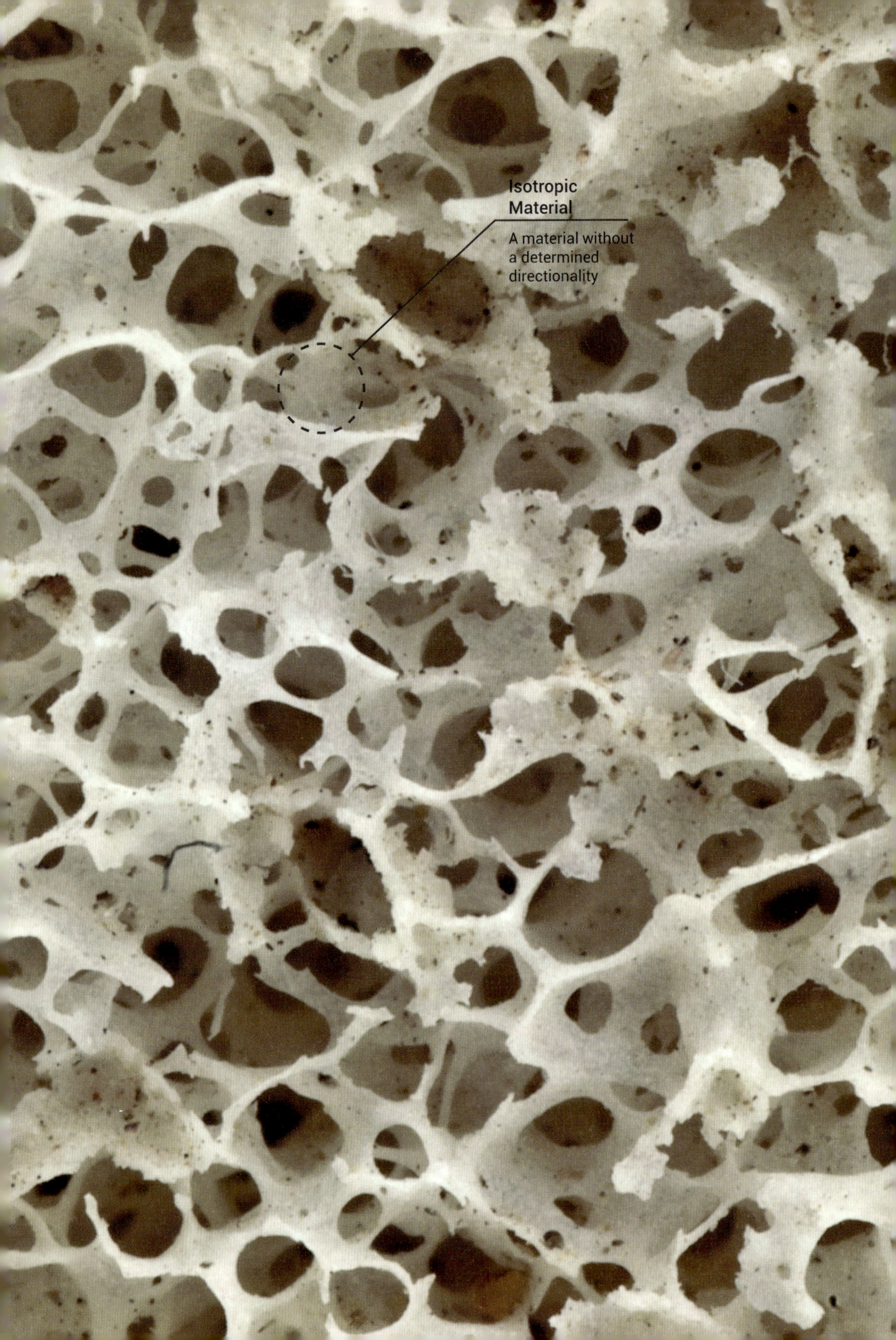
Isotropic Material
A material without a determined directionality

Isotropic & Anisotropic

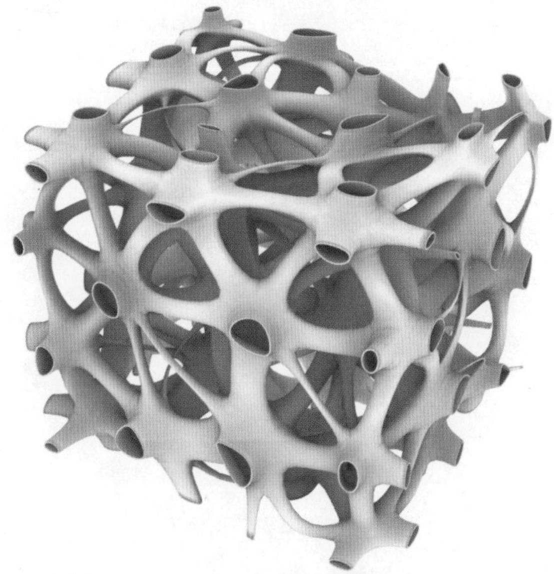

An object's or material's determined directionality is what comprises whether it is isotropic or anisotropic. This is important when studying properties, specifically those related to structural behavior. As an example, if a material tends to bend or move the same way regardless of the orientation of a force—this can be considered isotropic. If it has a specific directionality tending towards one way, for example, paper when ripped, this is assumed to be anisotropic.

Isotropic:

Exhibiting properties (such as velocity of light transmission) with the same values when measured along axes in all directions.[7]

Anisotropic:

Exhibiting properties with different values when measured in different directions.[8]

Material Directionality

Opposite Page:
Image of a macro cross-section of human hip bone with isotropic nature.

Current Page:
Conceptual diagram of an isotropic arrangement without directionality tending towards any axis.

---

7 "Definition of 'isotropic'." Take Heed/pay Heed Definition and Meaning | Collins English Dictionary. Accessed November 22, 2018. https://www.collinsdictionary.com/dictionary/english/isotropic.
8 "Anisotropic." Biology Online. Accessed November 22, 2018. https://www.biology-online.org/dictionary/Anisotropic.

Hierarchical Structures
Wood is one of the most common examples of structures that at different scales have a high ordered system

Hierarchical Structures

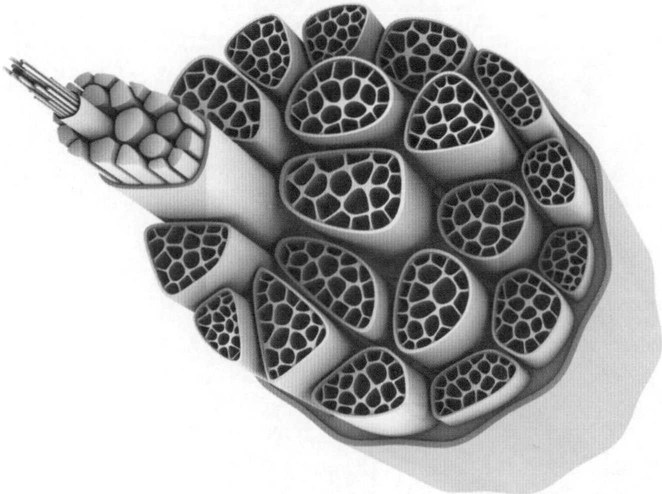

In many natural artifacts, there are systems of organization that have different scales of hierarchies. The most commonly known are those in bones and wood, which have systems that can be reduced down to the microscale that still hold to the hierarchy. These structures are useful to understand concerning organizational complexity and material properties.

Hierarchical Structures:

Biological organization is the hierarchy of complex biological structures and systems that define life using a reductionist approach.[9]

Reductive Structures

Opposite Page:
Image of woodcut showing pores and rings.

Current Page:
Concept diagram of structural hierarchy similar to that found in bones.

9 Solomon, Eldra P., Linda R. Berg, and Diana W. Martin. Biology. Belmont, CA: Brooks Cole/Cengage Learning, 2010.

Structural Color
Butterfly wing scales are commonly known for their ability to reflect light at certain frequencies

Structural Color

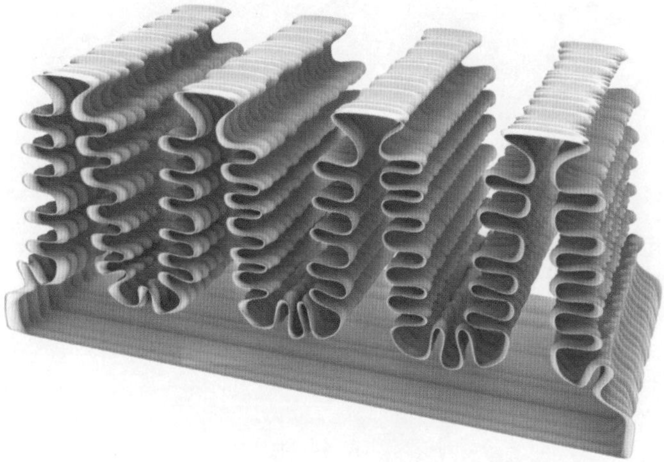

Structural color is an aspect known most commonly to the Blue Morpho butterfly and Peacock feathers. In both cases, the color of the wings or feathers is generated through reflected light at certain wavelengths which appears as color. The benefit of developing designs with structural color is that it eliminates most (if not all) of the pigmentation process, which when considering fashion or printing, is an expensive part of the process.

Structural Color:
A color or color component due to interference of light (as in thin films), diffraction (as by a grating), refractive dispersion (as in a rainbow), differential scattering, or differential polarization .[10]

10 "Structural Color." Merriam-Webster. Accessed November 22, 2018. https://www.merriam-webster.com/dictionary/structural color.

Reflective Color

Opposite Page:
SEM image of a Peacock butterfly wing.

Current Page:
Diagram showing conceptualized Blue Morpho butterfly scale detail which reflects light to appear blue.

Biological Prototypes

As a showcase of the current bio-inspired research going on in China, four renowned university labs were selected to introduce their work in this book: The Key Laboratory of Bionic Engineering at Jilin University, the Key Laboratory of Bioinspired Smart Interfacial Science and Technology at Beihang University, the Dynamic Technology Lab for Bionic Materials and Structures at Donghua University, and the Biomimetic Design Lab (BiDL), at Tongji University in Shanghai. The chosen projects cover research and applications ranging from material science to mechanical engineering and environmental design. All the projects are explained following an established methodology: a natural element is selected, some of its properties are highlighted and studied, real applications are explored and proposed, and eventually a final prototype is built and evaluated.

Plant Patterns

Small leaves arrayed on a plant

Jilin University

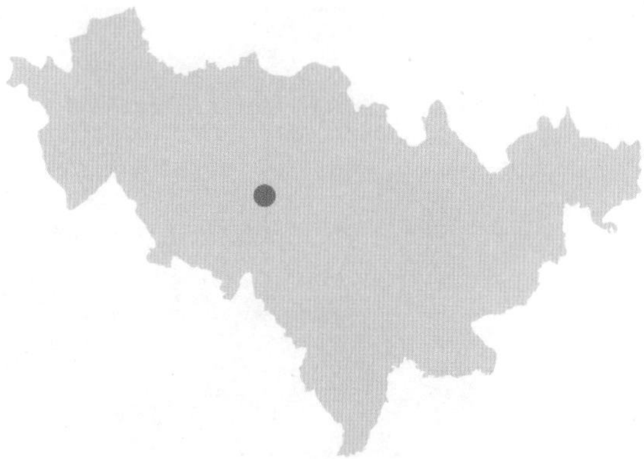

The Key Laboratory of Bionic Engineering of Jilin University was created in 2010 by Professor Zhiwu Han, and was the first laboratory supported by the Ministry of Education for research into biomimetic engineering. Its primary focus of study includes mechanical and molecular biomimetics, along with material and surface engineering. These topics span from design to material development and manufacturing.

In the last 5 years, from a funding total of 240 million RMB, the lab has undertaken more than 200 projects , publication of 15 books, registered 200 patents, and had 400 papers incorporated into the Science Citation Index (SCI). Currently, the lab's team consists of 60 people, including Professor Yan Liu, whose research group is responsible for the projects presented in this chapter.

Flowering Attractor

Opposite Page:
Image of Lily blooming with stigma and anther

Dynamic Curvature Radii

Earthworm

In the field of agricultural machinery, soil adhesion is a recurrent problem. Reducing soil adhesion on surfaces is one of the critical issues which can effectively improve the operational quality of agricultural machinery, especially in press rollers, soil adheres to the rubber bulge areas during use.

To explore this problem, we adopted a biomimetic approach to discover how nature addresses adhesion. Beginning with sub-terrestrial organisms, we found that many earthworms and insects live in soil without attracting dirt. We began our investigation by analyzing the surface features of worms. We first observed how the curvature radii changed into three states throughout movement: stretched, motionless and contracted. These movements were abstracted and used later in our prototypes.

The next observation showed that the surface layer of earthworms has a film that helps to reduce soil adhesion. Through extraction of the earthworm´s surface attributes, a basis for the design was established for the surface structure of a press roller. On these rollers, an elastic with six rubber bulges is typically used.

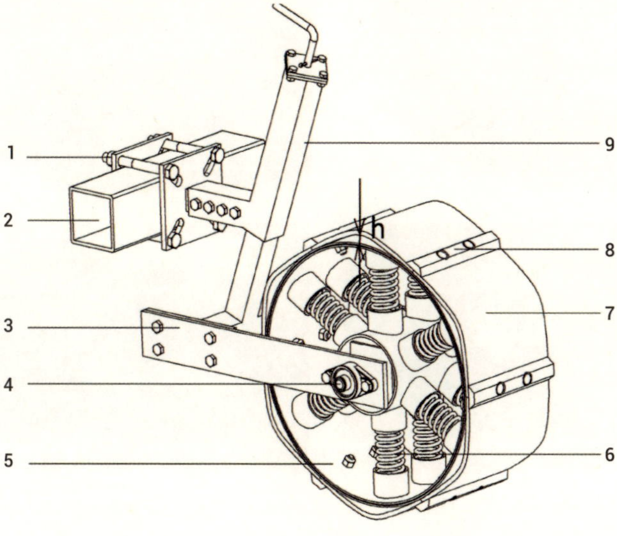

Earth Worms & Agriculture Machinery

Opposite Page:
Image of Earth Worm

Current Page:
The press roller diagram showing rubber bulge, the component studied for prototyping.

1. Grip plate
2. Rack
3. Roller frame
4. Centre axis
5. Cylinder shell
6. Elastic spoke
7. Rubber bulge
8. Rib
9. Adjustable height device

Research by Yan Liu.

Press Roller

Site test in an indoor soil bin. The trolley test system consists of rack, drive system, power output system, hydraulic system, control system and brake system set in soil that simulates field conditions.

Earthworm

Photo series depicting the surface features of earthworm's head and body at three states, including stretched, motionless, and contracted.

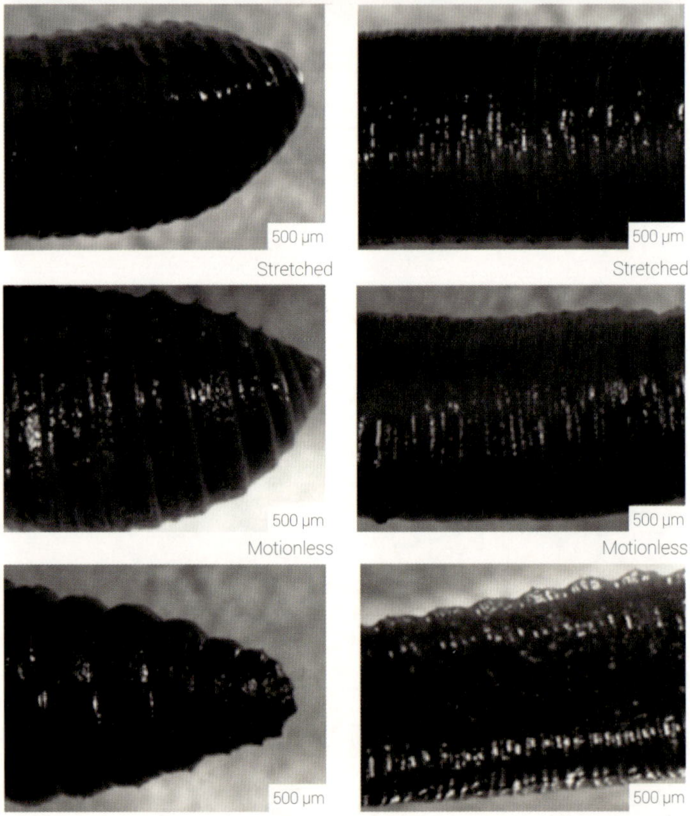

Stretched Stretched

Motionless Motionless

Contracted Contracted

(Head) (Body)

For the prototyping studies, each rubber bulge was developed to have a width-height ratio equal to that of the earthworm´s motionless state. From our studies, it was found that the primary factors affecting the soil adhesion on the surfaces of the press roller were the load applied and the height (h) of the rubber bulge. Therefore, the design of the rubber bulge was developed so that when there was contact with the soil, the rubber would be continuously squeezed through the rotation of the press roller. This made the curvature radius gradually decrease, thus reducing the overall contact area, and avoiding any adhered soil to break down.

From our testing showcased in the following page, we found the press roller prototype with the rubber bulge proportions h2 presented the smallest soil mass adherence.

In addition to the individual prototypes, a complete field study was carried out using a press roller. From this testing, the results showed that the press roller with the prototypical rubber bulge had a soil mass adherence 37.62% lower than the standard press roller. This results suggested that the experimental biomimetic approach explored in this study, would be a successful method for reducing the soil adhesion in agricultural machinery.

h1 and 0°

h1 and 10°

h1 and 20°

h2 and 0°

h2 and 10°

h2 and 20°

h3 and 0°

h3 and 10°

h3 and 20°

Infield Testing

Photos showing field test series. The range of curve h1 indicates the changes in adhesion force between the rubber bulge and surrounding soil as too small, which mostly restricts the reduction of soil adhesion.

Curve h2 represents the smallest soil mass adhered since it is most similar to the earthworm surface proportions. Thus, the press roller with rubber bulge h2 has the best anti-adhesion effect.

The range of curve h3 is too large, so the tension of rubber bulge is less than the maximum static friction on the contact interface, which continuously squeezed the rubber bulge and led to folding.

h1 and 30°

h1 and 40°

h1 and 50°

h2 and 30°

h2 and 40°

h2 and 50°

h3 and 30°

h3 and 40°

h3 and 50°

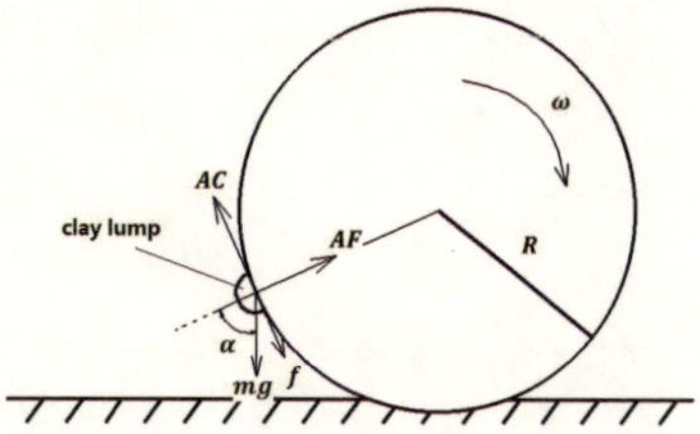

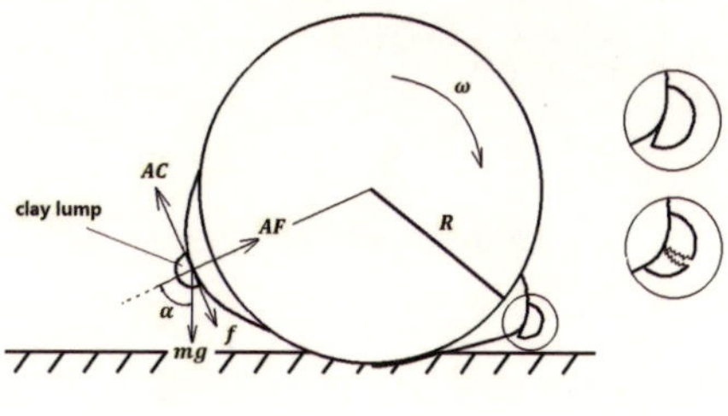

Adhesion Force

Diagrams showing conceptual relationships of clay adhesion to the surface of rubber bulge on a press roller.

Convex Domes

Dung Beetle

The dung beetle is typically a soil-burrowing animal and has adaptive features to fit their environment. They have a function of anti-adhesion and soil resistance which is attributed to the raised microstructures on their body surface.

The head of dung beetles is typically a surface covered with small convex domes. Through the use of a scanning electron microscope (SEM), we observed and analyzed the surfaces of several dung beetles. From these observations, we found that the raised domes on the surface are random in size, with a quantity that appears to be densely packed and uniformly distributed. Through abstraction of these relationships, we aimed to develop surfaces for machinery that would support anti-adhesion of the soil using similar properties.

Dung Beetle Domes

Opposite Page:
Image of Dung Beetle

Current Page:
SEM image of Dung Beetle shell surface displaying the convex dome surface conditions.

Research by Yan Liu.

Surface Design

Diagrams showing patterning options of convex circles intended to be applied to agricultural equipment surfaces. Images vary in slight size to explore surface friction conditions.

1

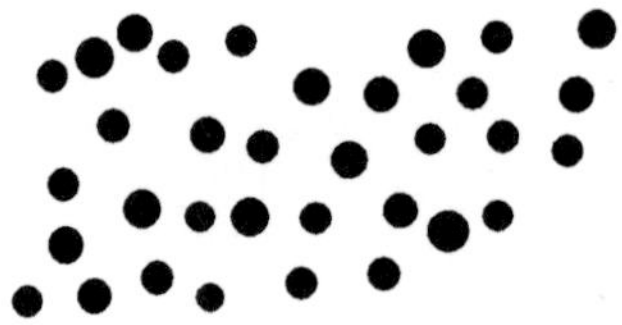

2

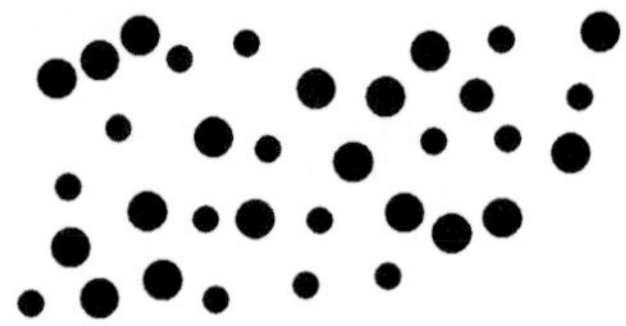

3

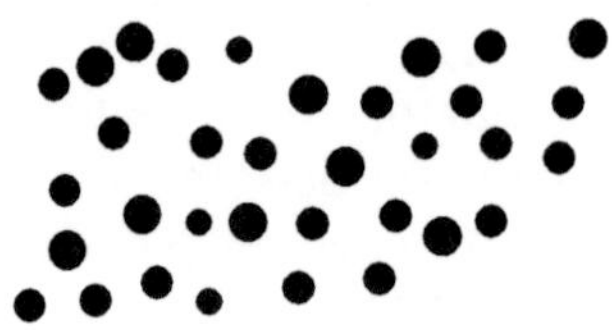

4

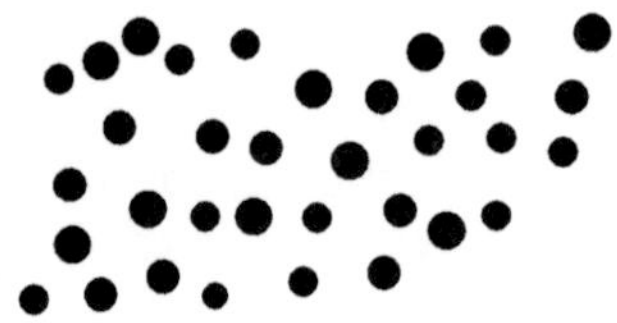

5

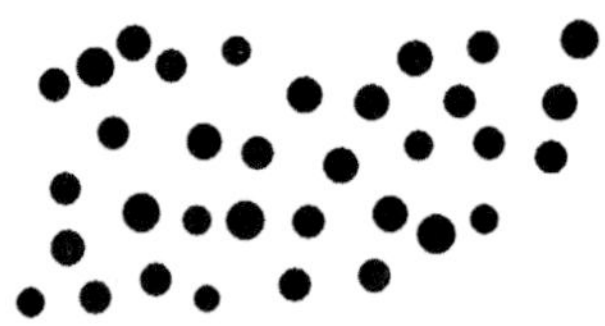

Surface Design

Computer generated surface showing raised convex pattern in 3D.

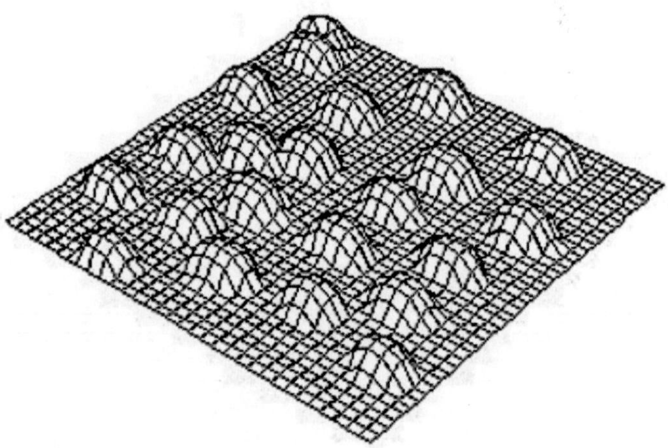

Through computer simulation and mapping of similar surface proportions through mathematical analysis, a biomimetic non-smooth surface was developed. The size of non-smooth units was simplified within manufacturing capabilities to adapt to the industrial production.

The application was for a plow moldboard which is a typical farming component in agricultural machinery. The biggest problem with the traditional plow moldboard is severe soil adhesion. Soil adhesion and friction on the surface usually generates a draft force where energy consumption and abrasion increase, which leads to the reduction in the production efficiency.

The parameters of the convex domes on the covered surface were at 7 mm in height, 25 mm in base diameter and 45 count. In the best case studied, the results showed that the machinery's resistance was reduced by 18.09% as compared to that of the conventional smooth blade.

From this testing results, we believe that the explored biomimetic method is one of the most effective means to reduce soil adhesion and resistance. The issue of anti-soil adhesion has been well developed in dung beetles and other soil-burrowing species (ant, pangolin, earthworm, ground beetle, mole cricket, roach, earwig, locust, millipede, and centipede) which all live in moist, soil-rich environments. These are all worth exploring due to the distinct non-smooth morphology of their body surfaces.

Plow Testing

Current Page:
Image Above-
Photo of standard plow moldboard with smooth surface.

Image Below-
Image of plow moldboards after field testing has occurred. The results show that in the main areas of the moldboard, there is little soil adhesion. The most accumulation occurs at the edges, where no surface treatment was explored.

Opposite Page:
Image of moldboards during field tests. Complete coverage of the moldboards is shown to indicate the standard exposure to soil.

Cactus Spines

Cactus

The cactus can survive in the desert due to its efficient water collection system. The clusters of needle-like spines or trichomes that grow on the stem's surface collect water during changes in temperature, primarily in the morning. The spine is composed of three parts with different structural features. The tip containing oriented barbs, the middle with gradient grooves and the base including belt-structured trichomes.

To study potential applications based on these needle-like spines, we developed tests to explore the relationship between form, surface material properties and physics of water collection.

One concept that facilitated the water collection of the spines was a feature called the wettability gradient. This occurs when a water droplet is driven by motion on a solid surface due to the surface's roughness, inclination, form and adjacent pressures pulling on the liquid.

Cactus Spines

Opposite Page: Image of cactus with clusters of spines that facilitates water collection.

Current Page: Close Up image of cactus spines.

Research by Yan Liu.

Due to the wettability gradient and gravity, the tip-shaped patterns of the spines can drive tiny water droplets towards more wettable regions. From this, we considered that an artificial surface with a tip-shaped pattern could potentially collect more water than other types of surface.

To test these theories, different water collection processes on various surfaces were explored. Compared with other samples, the test of the superhydrophobic and hydrophilic regions with the tip-shaped pattern produced unidirectional water transport on the surface as expected. The water droplets were driven by gravity and a wettability differential force (FWD) directed movement from the hydrophobic to hydrophilic tip. In the hydrophilic regions, with the increase of the water volume, the water moved from the tip to the bottom under the action of gravity (FG) and Laplace pressure (FL). This pressure was the result of the difference between the inside and the outside of a curved surface that formed the boundary between the liquid region.

These investigations may provide insights into designing bioinspired materials with interphase wettability for highly efficient water collection and unidirectional liquid transport, which is helpful to control the wettability of graphene and developments in fluidic nanodevices.

Directional Transport Testing

The photographs series displays 50-second elapsed testing of five control samples to explore the relationship between surface properties and directional conditions. Surfaces surrounding the tests were all hydrophobic.

1. Superhydrophilic Surface
2. Superhydrophilic Surface with Square-Shaped Pattern
3. Superhydrophilic Surface with Triangular-Shaped Pattern
4. Superhydrophilic Surface with Triangular-Shaped Etching
5. Superhydrophilic Surface with Circular Shaped Forms

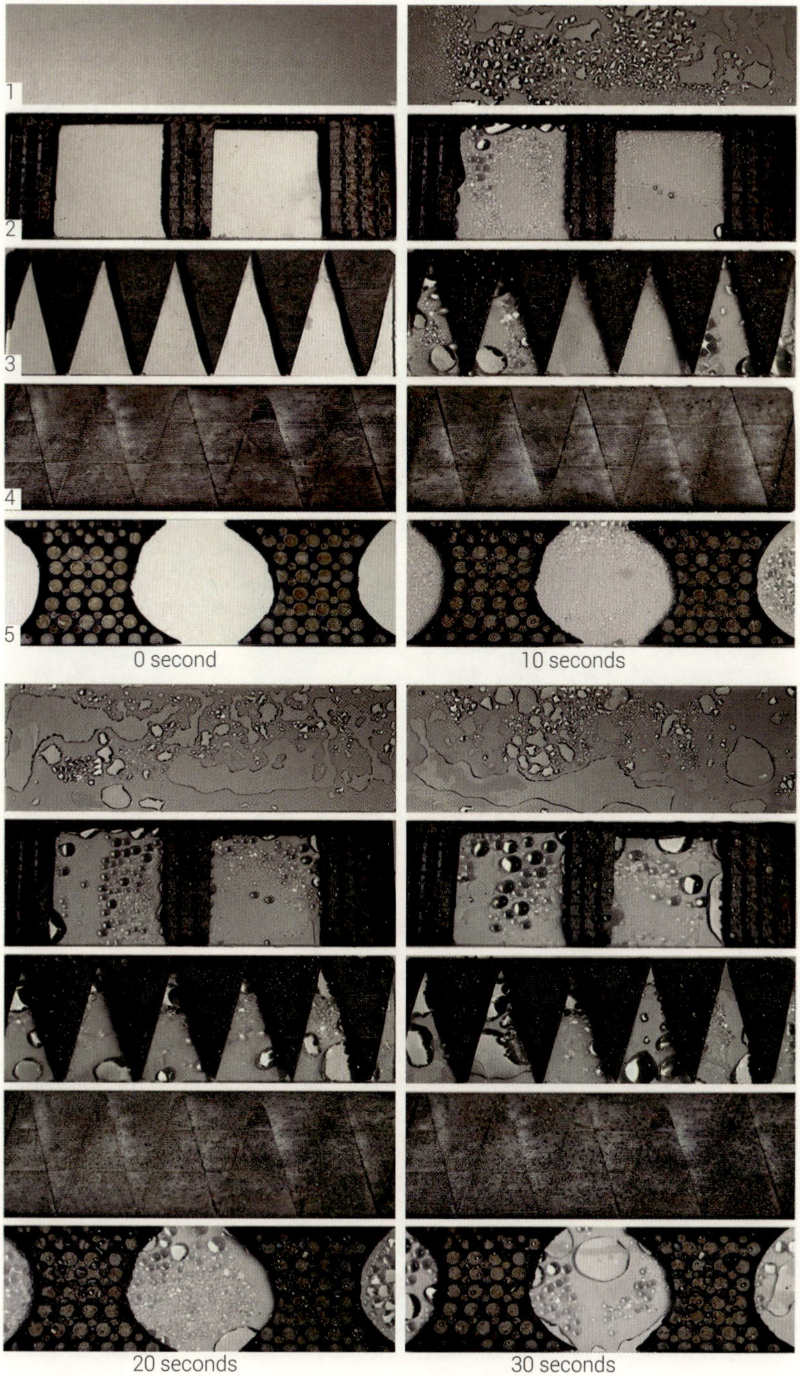
1
2
3
4
5
0 second
10 seconds
20 seconds
30 seconds

Microscale Papilla

Rose Petals

Rose petals are unique as they exhibit superhydrophobicity as well as high adhesion. The surface of a rose petal has a hierarchical structure consisting of microscale papillae with each papilla having nanoscale folds that contribute to these features. In the image below, both surface structures observed (micron-sized bumps, and submicron-sized folds) that increase the solid-liquid contact area.

To develop surfaces with similar properties, we established a process which combines graphene, copper and stainless steel into a panel. The process to produce this composite panel was explored by growing graphene on a copper substrate where the copper grains were clearly visible. Micro/nano structures were etched by laser and combined onto the graphene film. The end result showed similar properties to rose petal structures.

Microscale Papilla

Opposite Page:
Image of rose petals with water droplets collected on the surface, an example of the petal's superhydrophobicity.

Current Page:
Microscale view of rose petal surface with papillae that contribute to the petal's superhydrophobic and high adhesion features.

Research by Yan Liu.

Petal-like Structures

Images:
Composite surface steps of fabricated bioinspired rose petal surface.

Laser etched "horseshoe" shaped structures and micro bumps on steel.

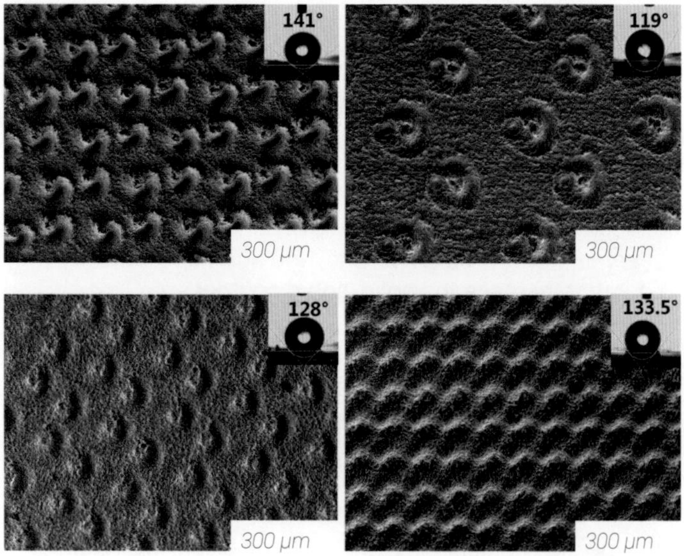

Petal-like Structures

Images show graphene that was grown on copper with visible copper grains.

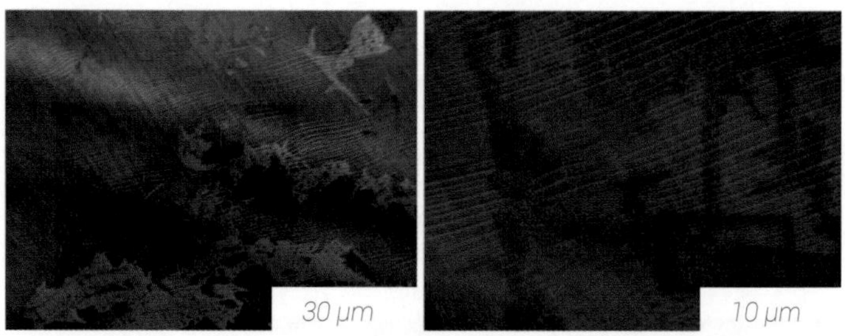

Petal-like Structures

Final graphene films with superhydrophobicity and high adhesion successfully transferred on to stainless steel.

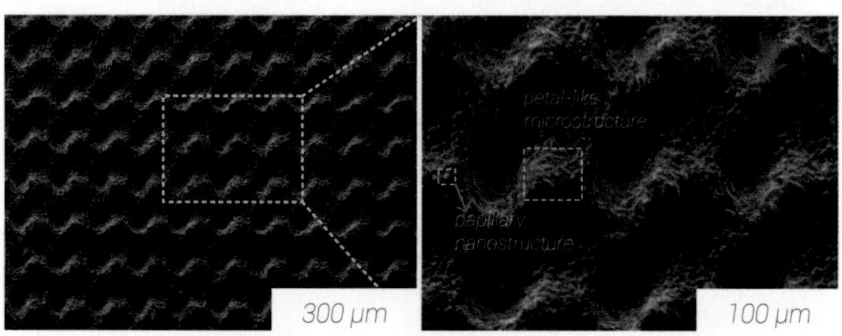

Super Hydrophobicity

Artificial material displaying super hydrophobic properties. The super-hydrophobic attributes could be sustained for nearly a month. A water droplet could pin to the graphene surface firmly no matter the angle even when the surface was turned upside down.

High Adhesion

Artificial material displaying high adhesion properties. To transfer micro-droplets successfully, the liquid transportation process shown appears potentially suitable for chemical reactions and similar applications. The composite interface is promising for the design and construction of smart interface materials and microfluidic devices.

SEED
AWN
AWN NEEDLE

Wheat Awn

Wheat possesses the ability to move and bury its seeds. This movement is activated by humidity, which constricts fibers, causing rotation. This movement is useful for seed dispersion, as once the awn seed has fallen on to the surface of the soil, it becomes embedded into the ground through this motion.

From observation, we noted that the awn exhibited a gradient microstructure in pore size, which plays an essential role in the ability of the plant to rotate. This showed that through varied placement of pores, different movements could be programmed into a shape.

The concept of this research was to explore movements by integrating hydrogels with graphene. Informed by the microstructure of the awns, a thermo-sensitive hydrogel with microstructure was developed that followed similar proportions to awns. The first step was to achieve a resultant form after water was fully absorbed by the hydrogel. This absorption provided the material base for loading graphene particles into the hydrogel.

The aim was to develop a hydrogel to exhibit flexing motions of a real hand. Under the concentration of multiple laser spots, the hydrogel hand could realize gestures of "OK", "victory" and "fist". As the intent of the experiment was to look at potential developments of form activation through stimuli, the prototype was deemed successful. The concept of contraction and force from internal pressures were the beneficial aspects found in the wheat awn structure.

Wild Wheat

Opposite Page: Image of wild wheat with awns.

Morphology of wild oats seed: Seed, Awn, Awn Needle

Research by Yan Liu.

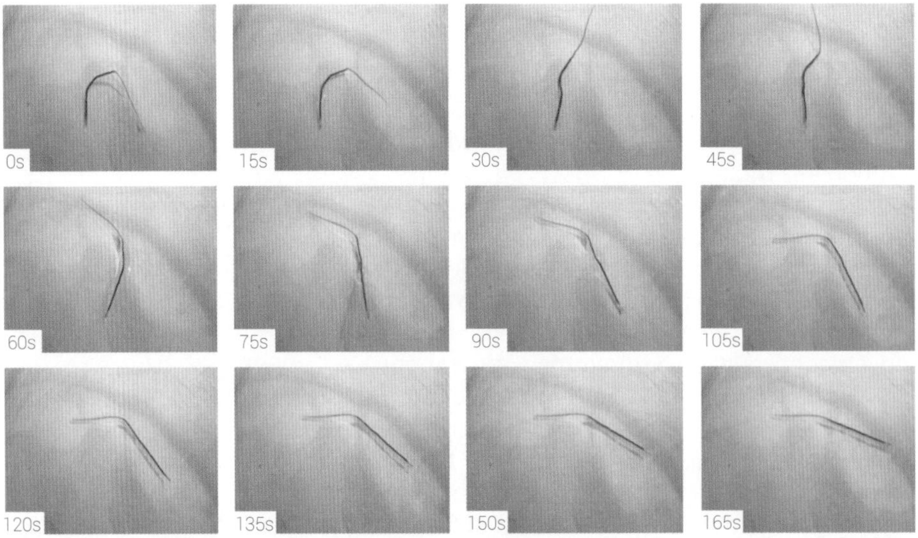
0s
15s
30s
45s
60s
75s
90s
105s
120s
135s
150s
165s

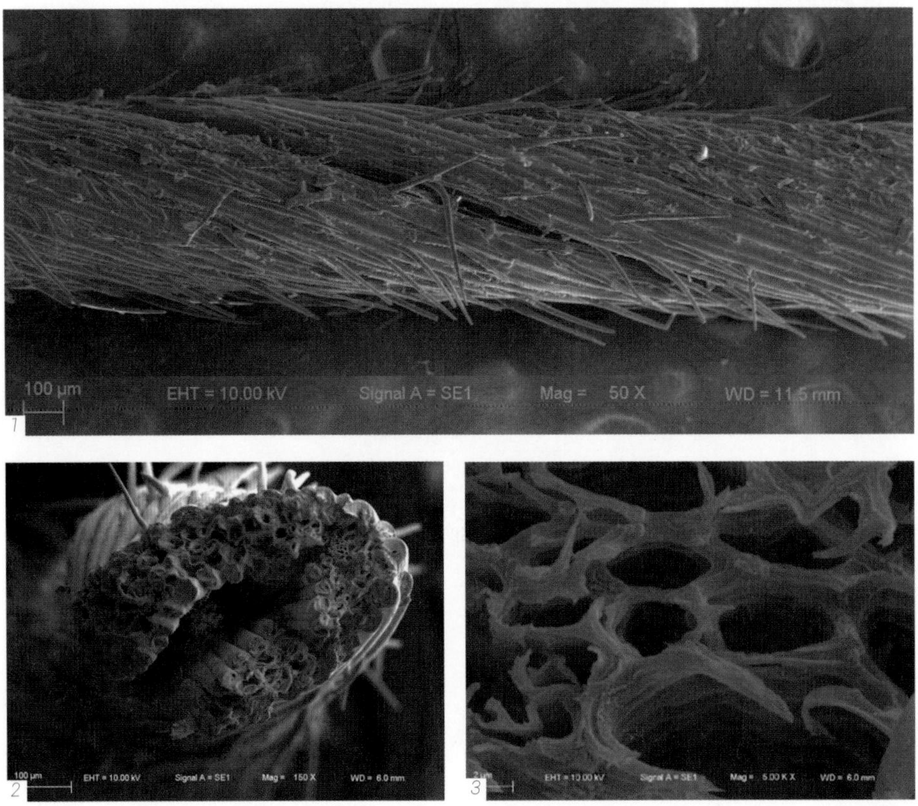
100 µm
EHT = 10.00 kV
Signal A = SE1
Mag = 50 X
WD = 11.5 mm
1
100 µm
EHT = 10.00 kV
Signal A = SE1
Mag = 150 X
WD = 6.0 mm
2
2 µm
EHT = 10.00 kV
Signal A = SE1
Mag = 5.00 K X
WD = 6.0 mm
3

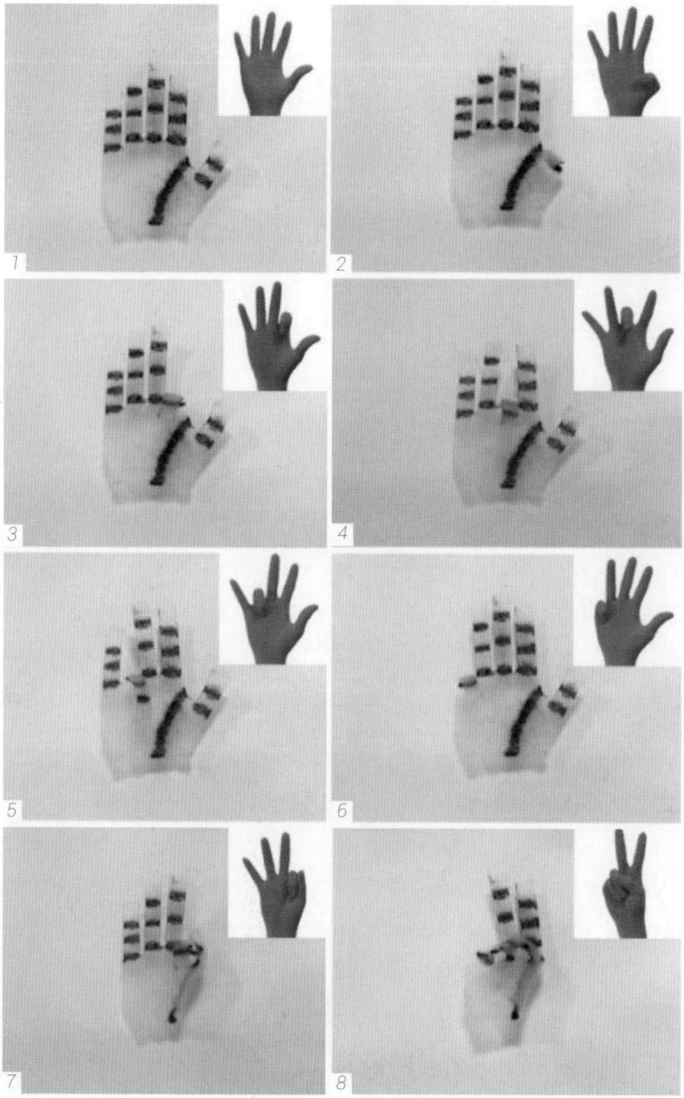

Hydro-gel Hand

1. Initial state of the hand
2. Deformation of the thumb
3. Deformation of the index finger
4. Deformation of the middle finger
5. Deformation of the ring finger
6. Deformation of the little finger
7. Gesture of "OK"
8. Gesture of "victory"

Wheat Awn

Opposite Page:
Top Image-
The torsion deformation process of wild wheat in water.

Lower Images-
1. Spiral morphology
2. Cross section micro-structure
3. Gradient pore characteristic of wild oats

Beihang University

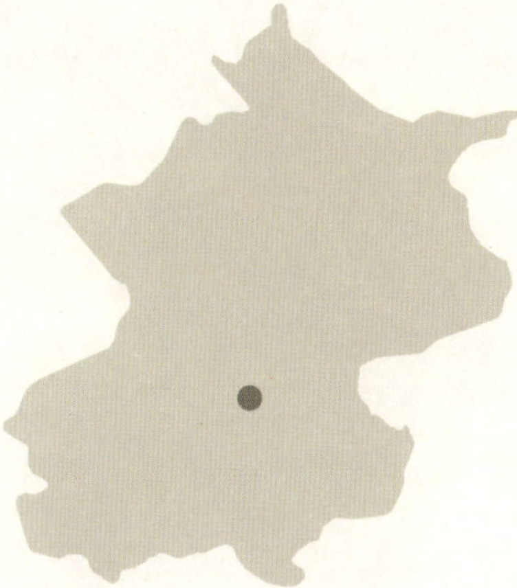

The Key Laboratory of Bioinspired Smart Interfacial Science and Technology was established by Prof. Lei Jiang in 2009, as part of the School of Chemistry at Beihang University. The lab's research theme identifies unique attributes of vegetable and animal tissue for the development of "smart materials". Smart materials are engineered to have properties that can be modified in response to external stimuli, such as temperature, light or electric fields. These new methods for the design of bio-inspired interfacial materials are on scales down to the micro and nanometer levels. The projects included are developed under the lead of Dr. Yongmei Zheng and Dr Lei Jiang. Zheng's research specializes in bioinspired surfaces to control dynamic wettability, while Jiang's research field is the development of bioinspired smart multiscale interfacial materials (BSMI).

Leaf Patterning

Opposite Page:
Image of leaf veins

Spider Silk

Spider silk, which is composed of humidity-sensitive hydrophilic flagelliform (slender, long and flexible proteins), has one of the strongest reputations as a fiber with excellent mechanical properties. It is seen in webs built by the Cribellate spiders, also known as Orb spiders, which use a cribellum—a comb-like device—to separate silk fibers drawn from their spinnerets into many exceedingly fine fibers. Puffs composed of nanofibrils are spaced along two main-axis fibers with different arrangements. The puffs are known to be formed of random nanofibrils and show a variety of patterns.

In addition to this intriguing feature of the Orb spiders, however, one less studied, is its ability to collect water from the air. These highly hydrophilic nanofibrils in spider silk enhance the wettability (the tendency of one fluid to spread on or adhere to) spider silk, which is favorable for condensing water drops. When dry spider silk is placed in humid air, its structure changes as water starts to condense and form drops. This displays the ability of water collection in the air.

In our research, we have found that bioinspired spider silks can be designed into multi-functions. The silk can collect water through a variety of different orientations and designs by applying similar relationships of such hydrophilic nanofibrils. They can be used to enhance the already extensive applications for collecting water in the engineering fog-harvesting field.

Hydrophilic Nanofibrils

Opposite Page:
Spider web with water droplets.

Current Page:
Water droplets condensing along hydrophilic nanofibrils, as spider silk enhances the tendency of fluids to adhere to the web.

Research by Yongmei Zheng.

Structural Character of Spider Silk

SEM images show that dry spider silk is composed of two distinct periodically repeating regions termed "puffs" and "joints".

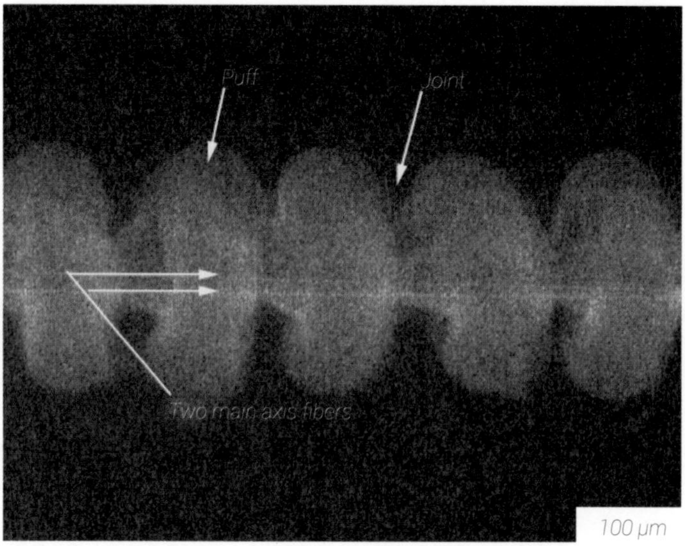

Spider Silk Nanofibrils

Magnified images showed that the puffs were composed of randomly oriented nanofibrils. These fibers are known to be highly hydrophilic and provide high wettability of the spider silk.

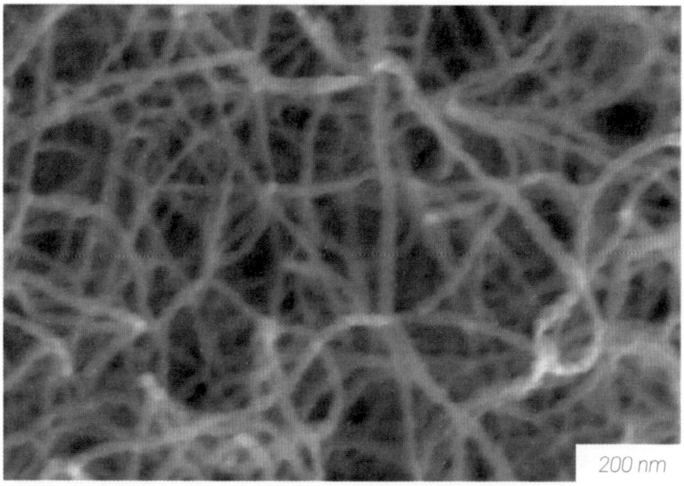

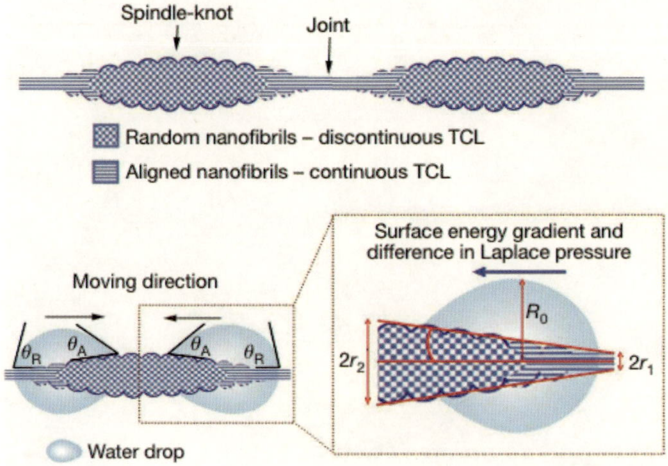

Water Collection

Diagram showing concepts of surface energy gradients from variations in roughness are assumed to drive water to collect at the more hydrophilic spindles, even when the drops are formed initially at the joints.

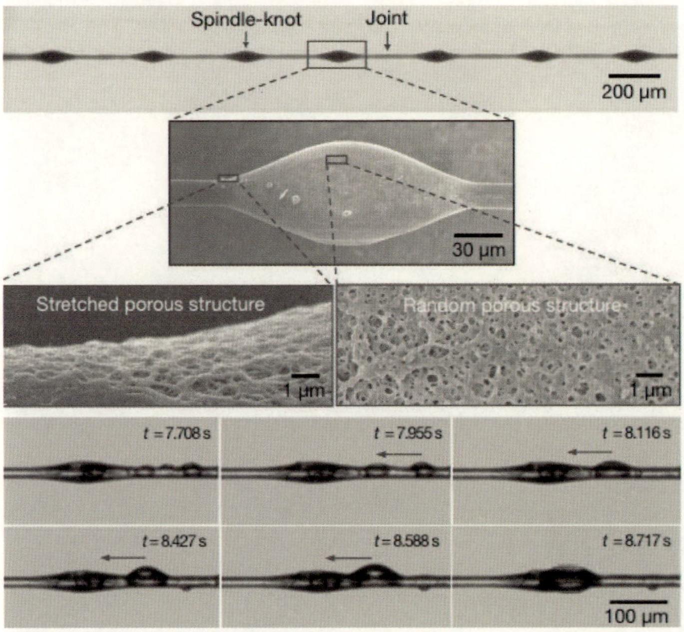

Porous Structure

The images show the lab's artificially generated spider silk exhibiting the same behavior of the natural fibers. These design principles could be used in the study of fiber development for water and aerosol collection in manufacturing.

Scales

Butterfly Wings

The tropical butterflies from Central and South America have very fine scales on the surface of their wings which are known for their ability to reflect light which appears as color. The Blue Morpho butterfly is one of the prime examples for its structural color. This property also keeps the wings clean and dry due to the structure of their scales. Similar concepts could be applied to self-cleaning fabric coatings.

Different from the self-cleaning property of lotus leaves where dust can be washed away from any direction, water drops falling onto butterfly wings can only move from a single direction away from the body.

When the butterfly wings slope downward, the scales and rear separate from each other, fragmenting the water contact line, making the water drops roll off quickly.

As part of our research, we developed a material that would be able to use directionality to its benefit. Through our prototypes, directionality was explored when activated by magnetic charge. We developed a surface fabrication method through magnetic micropillar arrays to achieve similar properties to the scales wettability. Negative PDMS (Polydimethylsiloxane) arrays are obtained by soft molding onto silicon micropillar arrays. Such developments could be applied for materials or surfaces that would benefit from easy-cleaning coatings.

Anisotropic Wettability

Opposite Page: Image of Morpho Butterflies

Research by Yongmei Zheng.

PDMS Soft Mold

Casting PDMS/NdFeB Composite

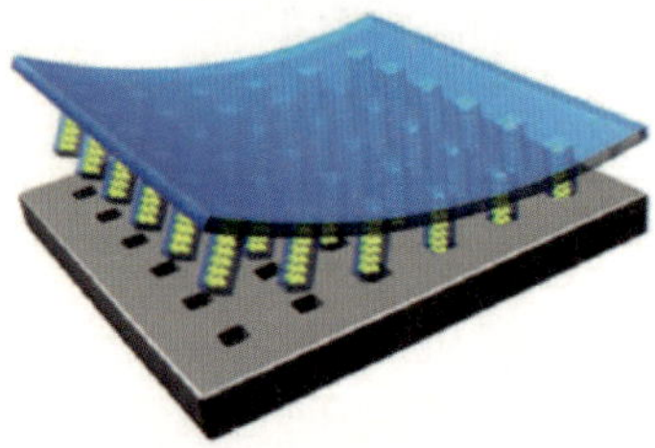

De-molding

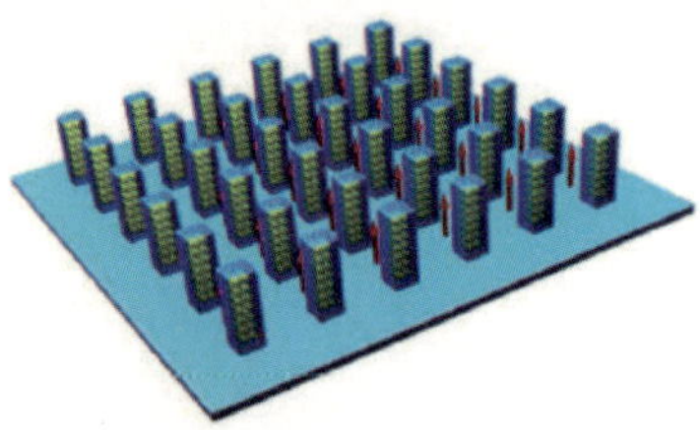

Dip coating

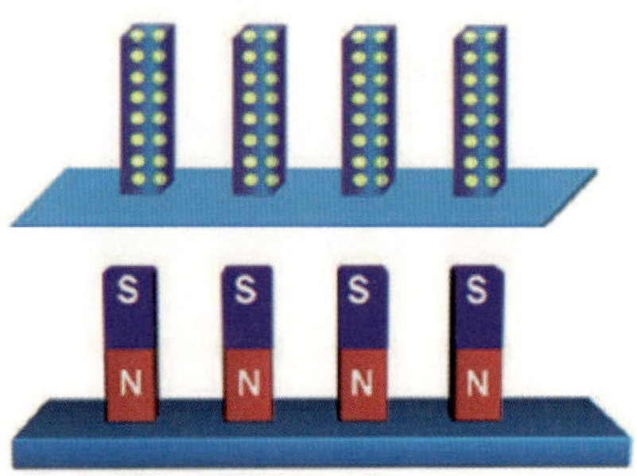

Magnetization

Controlled Directionality

Process of surface fabrication of magnetic micropillar arrays

1- casting PDMS/NdFeB composites and curing
2- demolding, dip coating, and silanization
3- magnetization along one direction
4- after the micropillars are magnetized, each micropillar is a micromagnet with the south (S) pole on the top and the north (N) pole at the bottom as the magnetization direction.

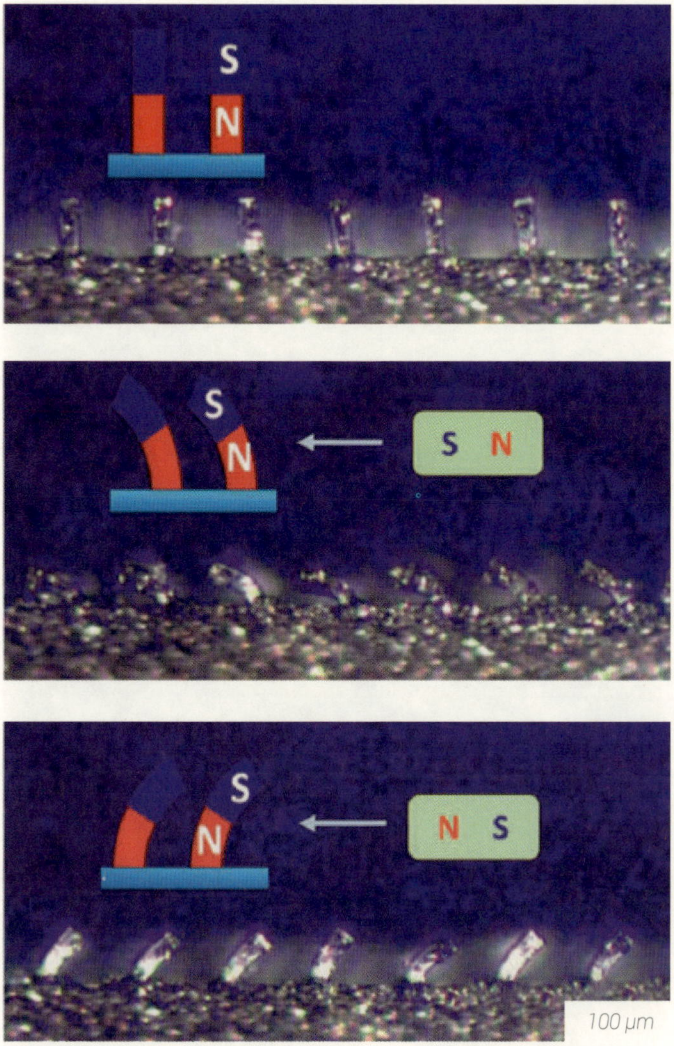

Material Prototypes

Images showing prototype directionality when activated. Such developments could be applied for materials or surfaces that would benefit from easy-cleaning coatings.

Papillae

Lotus Leaves

Lotus leaves have self-cleaning properties that allow them to remove dust from their surfaces. This feature is attributed to the hydrophobic attributes of the micro- and nanostructures of the plant surface.

The structures of lotus leaves can be artificially fabricated by soft lithography and nanotechnology to provide similar features. As part of our research, we explored surface water repellency at a low temperature which displayed anti-icing properties suitable for further material exploration.

We developed a biomimetic surface that could be flexible and superhydrophobic composite materials, namely PDMS (silicone polymer) and zinc oxide (ZnO). The surface showed water and ice repellency at low temperatures (-20°C) and at high relative humidity of 90% for up to three months.

The features of the surface come from the integrated effect of the composite flexibility and the multi-structure composed of PDMS micro papillae and zinc nano hairs. The zinc hairs cause the repellency of condensed droplets, whereas the PDMS microstructures are responsible for the suspension of droplets resulting from the coalescence of tiny condensed droplets. The composite structures developed showed stability and long-term water and ice repellency.

This study offers insights into the possibility of multi-structured materials which could contribute towards the development of robust anti-icing and deicing surfaces for practical applications.

Lotus Leaf

Opposite Page:
Image of lotus leaf with water beads.

Research by Yongmei Zheng.

Frozen Papillae

Images showing the surface reactions to icing at a low temperature and high humidity.

Papillae Reaction

Diagrams showing the conceptual reaction of water droplets to papillae at low temperatures with high humidity.

Red = High Pressure
Blue = Low Pressure

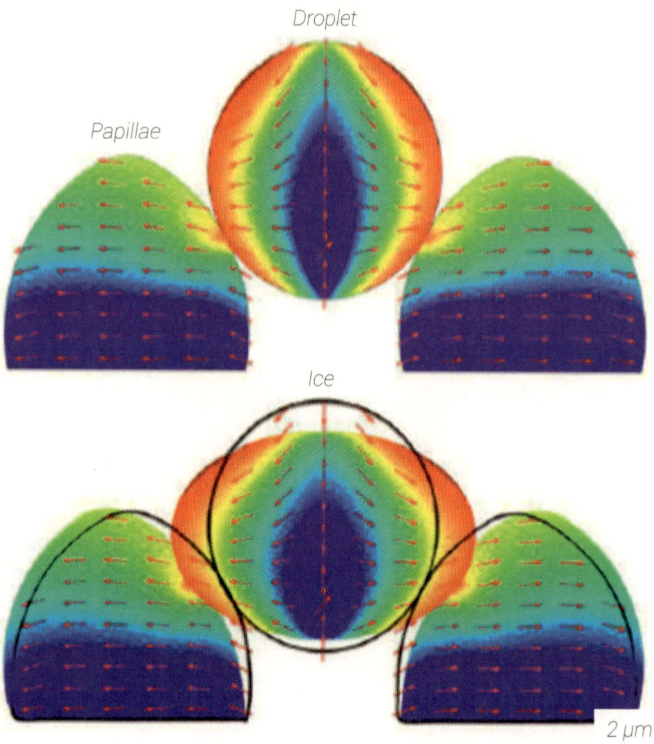

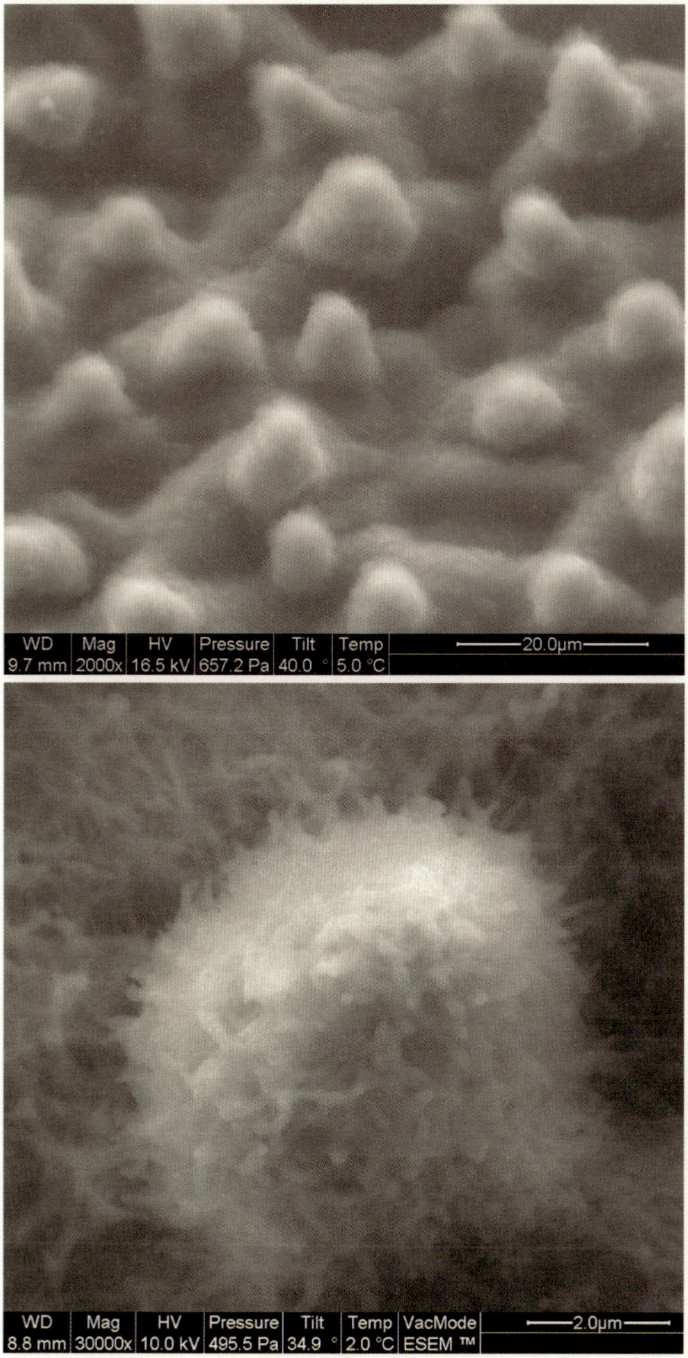

Lotus Structures

Detailed images of distributed papillae along the surface of the lotus leaf after exposed to cold temperatures and high humidity.

Microspheres

Mosquitoes Compound Eye

A mosquito's compound eye has inherent anti-fogging properties to allow it to thrive in its natural environment, close to water sources. This property is a consequence of the microscale and nanoscale structures of the eye's surface, capable of repelling water molecules.

Our research has developed a strong anti-fogging film, which is formed from hollow microspheres and ZnO nanorods that generate the same effect as the mosquito's eye. The surface is fabricated via combining airless spray and crystal growth methods. The microspheres strengthen the superhydrophobic qualities of the micro and nanostructure of the film, so that the anti-fogging effect is even possible at low temperatures.

When the film is placed in a low-temperature, high-humidity environment, supercooled tiny water droplets cannot spread. This investigation provides insight into the design of structured thermal insulation surface materials for anti-fogging that could be applied towards micro-devices used in cold, high-humidity environments.

Microspheres

Opposite Page:
Image of a mosquito in wet environment.

Current Page:
Microscale view of microspheres on the compound eye of a mosquito.

Research by Yongmei Zheng.

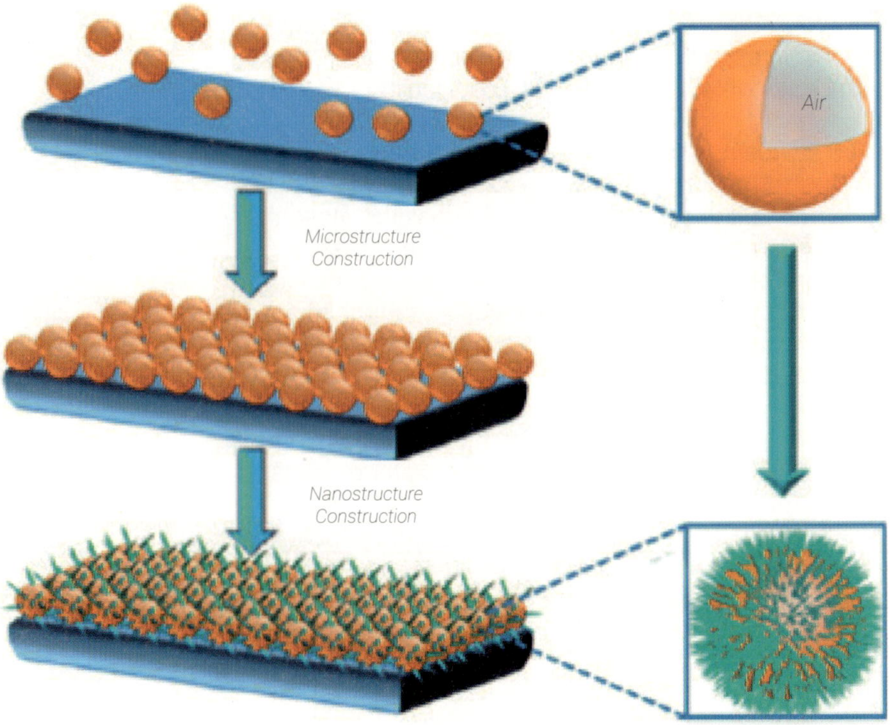

Superhydrophobic & Anti-fogging

Diagram showing surface testing. When the film is placed in a low-temperature, high-humidity environment, supercooled tiny water droplets cannot spread on it and can easily coalesce.

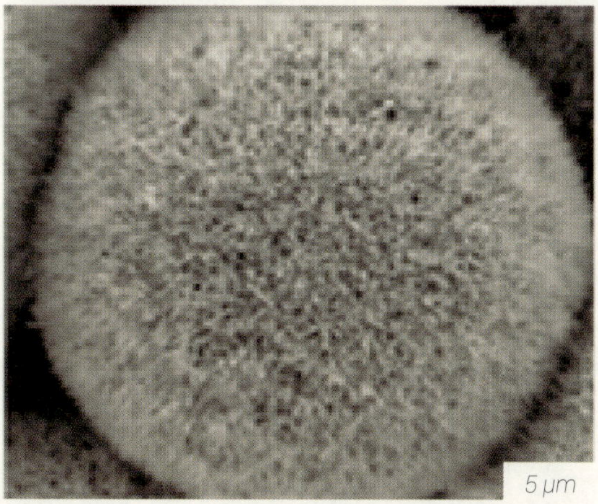

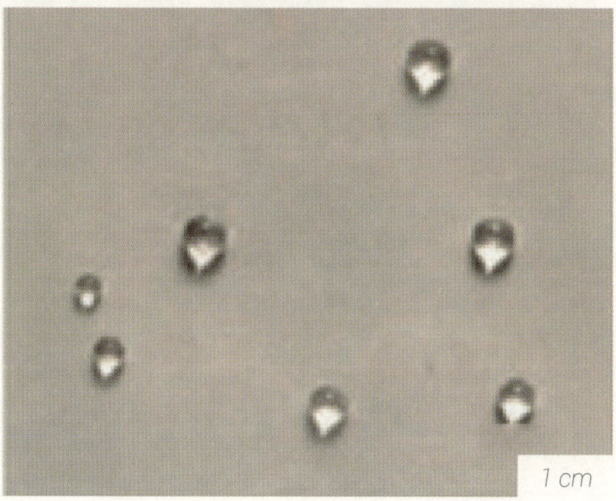

Anti-fogging Film

Top Image-
Artificial microspheres adhered to the surface.

Middle Image-
Water reaction in high-humidity, low-temperature environment.

Lower Image-
Artificial film reaction to water droplets showing surface's superhydrophobic properties.

Cactus Spines

Cactus

Cactus spines are known for their ability to collect water from the air during the morning periods using their thorns, called spines. The spine is a rough conical geometry, which due to this shape, generates a Laplace pressure between water droplets on spine. This effect causes droplets to move along the spine from the tip to end, which brings water closer to the cactus plant.

The *Gymnocalycium Baldianum* cactus, or commonly known as the spider-cactus or dwarf chin cactus, has clusters of 5–6 spines along every nodule. The physical characteristics of the spine are organized in such a way that there is a hierarchical system of water transport. Due to the cavities between splayed scales on the cactus and the grooved base of spine, the water droplets can remain as an ultra-thin liquid film along overlapping scales to form the ultra low-adhesive surface on the spine.

To explore this type of water collection, we developed a prototype to harvest fog using similar surface relationships. Through the use of a conical copper wire, named PCCW, we could collect fog on points of the conical surface from the air. Each point could then transport the droplets without external force. It was found that the ability of fog collection could be related to varied angles of the wires. Consequently, we developed a fog collector with an array system of PCCW designed to achieve a continuous process of efficient water collection.

Spines & Water

Opposite Page:
Image of cactus with spines along ridges.

Research by Yongmei Zheng.

Cactus Features

Current Page:
Image Series-
Water droplet moving along the cactus spine, from the top towards the base.

Capillary Features

Opposite Page:
Top Images-
The illustration shows the scale-covered spine. The space between the scale and the surface of the spine forms a splayed capillary channel, opening different radius at the two ends. Under the scale, there are parallel grooves that assist in directionality.

Lower Images-
Diagram showing physical relationships which allows for the conical capillaries to move water.

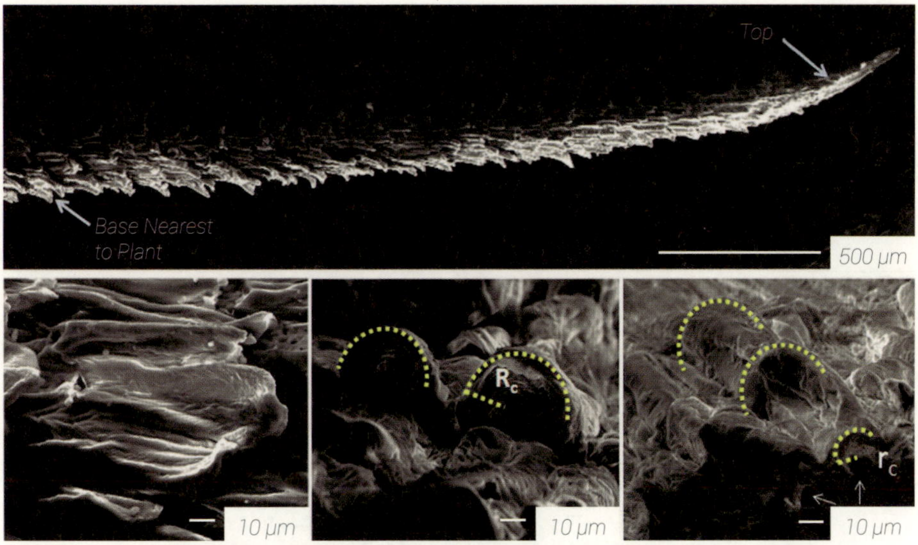
Top
Base Nearest
to Plant
500 μm
Rc
rc
10 μm
10 μm
10 μm
Scales
Conical Capillaries
Scale Cavities

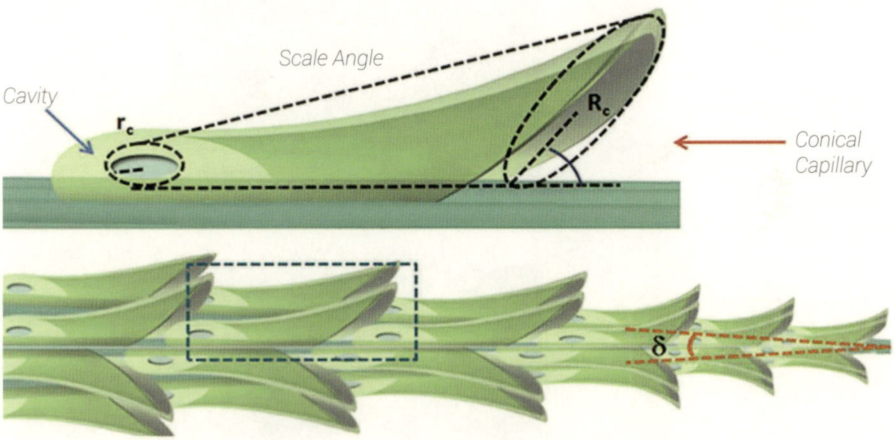
Scale Angle
Cavity
rc
Rc
Conical
Capillary
δ
Scaling of Cavities

Roughness Gradients

Current Page:
An artificial periodic roughness-gradient was fabricated using a conical copper wire (named as PCCW) inspired by the cactus spines.

Top Image-
Surface conditions near the tip of the spine

Middle Image-
Surface conditions near the middle of the spine

Lower Image-
Surface conditions near the base of the spine

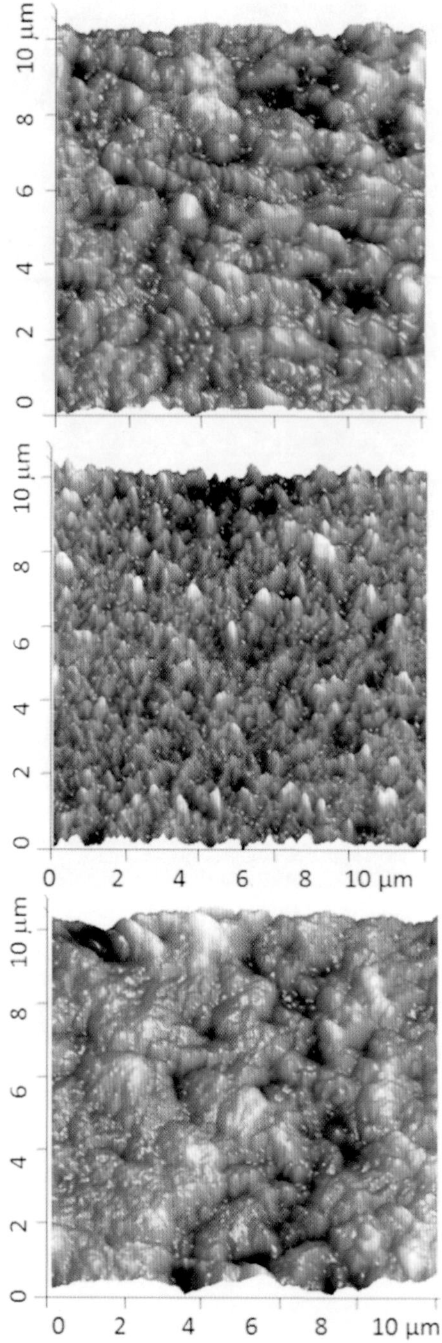

Capillary Prototype

Opposite Page:
Top Images-
Diagram of splayed capillary micro-tubes which induce self-transport of the drop. Drop moving continuously on the spine with splayed capillary micro-tubes composed of overlapping titled-up scales from the top to the bottom of the spine to facility self-watering.

Lower Images-
Diagrams showing the concept of PCCW prototype to harvest fog.

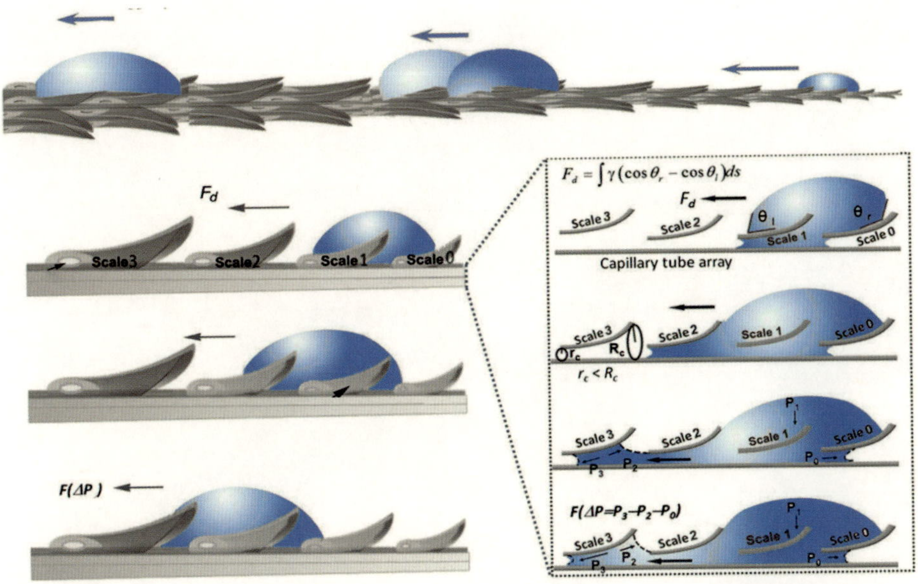
F_d
Scale3
Scale2
Scale1
Scale0
$F(\Delta P)$
$F_d = \int \gamma(\cos\theta_r - \cos\theta_i)ds$
Capillary tube array
$r_c < R_c$
$F(\Delta P = P_3 - P_2 - P_0)$

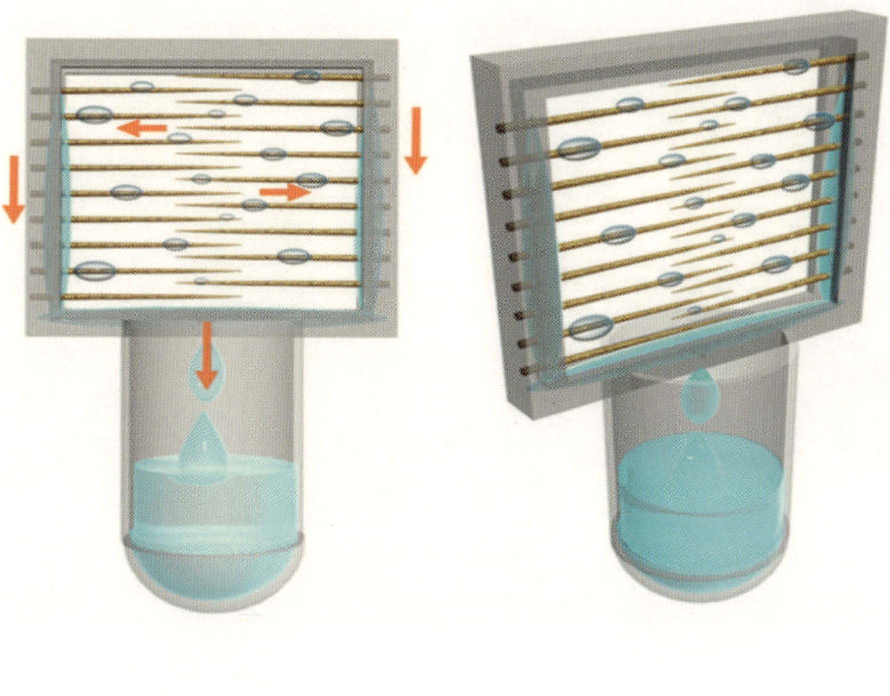

Water Repellent Legs

Water Strider's Legs

Water striders are insects known as pleuston, which are organisms whose habitat is the surface of a body of water. This is possible thanks to their remarkable water-repellent legs that enable them to stand effortlessly and move quickly on the water. It is the particular hierarchical structure of the legs, which are covered by large numbers of oriented tiny hairs (microsetae) with fine nano grooves, that are the key to inducing this water resistance.

On the legs of a water strider, there are thousands of oriented setae, which are needle-like elements, with diameters ranging from 3 micrometers down to several hundred nanometers. Most setae are roughly 50 µm in length and arranged at an inclined angle of about 20° from the surface of the leg. Many elaborate, nanoscale grooves are evident on each microseta, and they form a unique system of hierarchical structures.

This hierarchical micro- and nano-structuring on the leg's surface is responsible for the hydro- resistance that allows these insects to support themselves on a water layer, even if they are being bombarded by raindrops. This discovery is helpful in the design of miniature aquatic devices and water repellent materials for marine or wet environments.

Water Resistant

Opposite Page:
Image of water striders on top of the surface of the water. It was thought the bugs "trap air" to help them float, when it actually is the micro-structures of hairs that allow them to sit on the water surface.

Research by Lei Jiang.

Needle-like Hairs

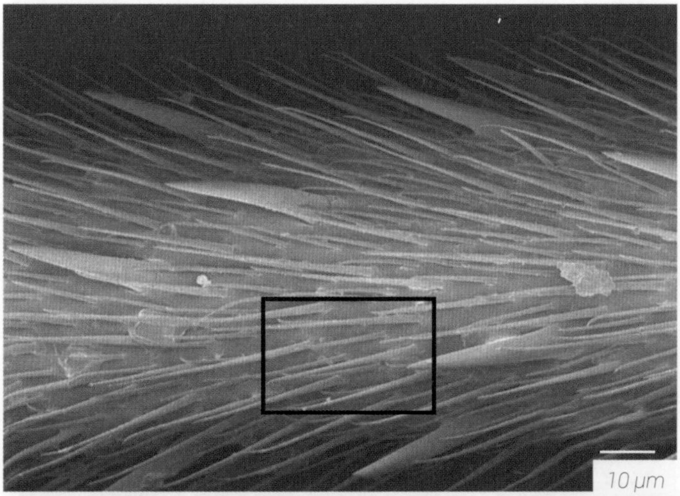

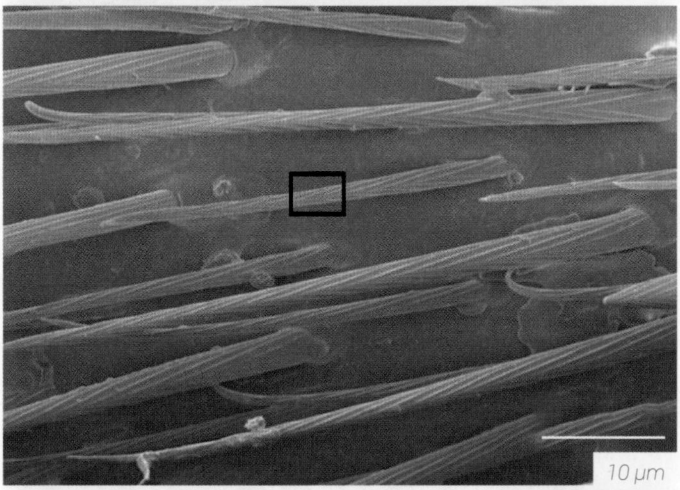

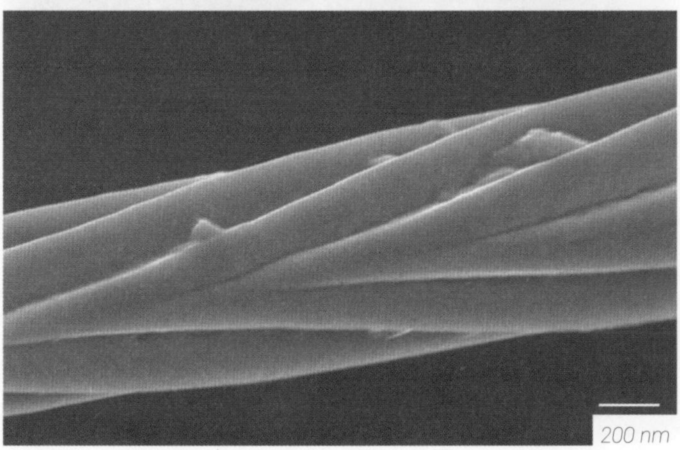

Hierarchical Structures

Current Page:
Top Image-
It shows the setae which area diameter of about 3 micrometers.

Middle Image-
Setae in greater detail showing arrangement and length.

Lower Image-
One seta showing detailed groves of needle-like hair.

Opposite Page:
Top view of water strider on water surface.

Needle-like Clusters

Cactus Clusters

The cactus *Opuntia microdaisys*, which originates from the Chihuahua Desert, has an integrated multi functional system that facilitates efficient fog collection. Clusters of needle-like spines and trichomes grow as a well-distributed array on the stem's surface with distances ranging from -7 to 23 mm. These spines grow in a range of directions with an average angle of 18.1±5.3° between the nearest two spines, forming a hemispherical structure.

The spine is composed of three parts with different structural features, the tip contains oriented barbs, the middle contains gradient grooves, and the base contains belt-structured trichomes.

The magnified image of the barb on the next page reveals a conical shape (apex angle 2β: 19.5±3.3°) with aligned grooves. The magnified images of the middle portion of the spine reveal multi-level grooves. The first-level grooves (the epidermal cells of the cactus spines) are microgrooves and have a gradient in width along the spine. The second-level grooves (the folding sculptures of the epidermal cell) are primarily sub-micro grooves lacking a noticeable gradient in width. As a result, the deposited water drops are driven from the tip to the base of the spine due to these relationships.

The subtle integration of these structures (conical spines and barbs, oriented barbs, gradient grooves and belt-structured trichomes) contributes to the superior ability of this natural mechanism to capture water droplets from the fog. Further investigation into the structure-function relationship of this system may provide insights into new materials and devices for efficient water collection from the fog.

Needle-like Clusters

Opposite Page: Image of *Opuntia Microdasys* Cactus known as "Bunny Ears" Cactus

Research by Lei Jiang.

Clusters

Detailed image of a cluster with water droplets collected at tips.

Water Transport

Diagrams showing how spine form and micro surface features organize water flow direction towards base nearest to cactus.

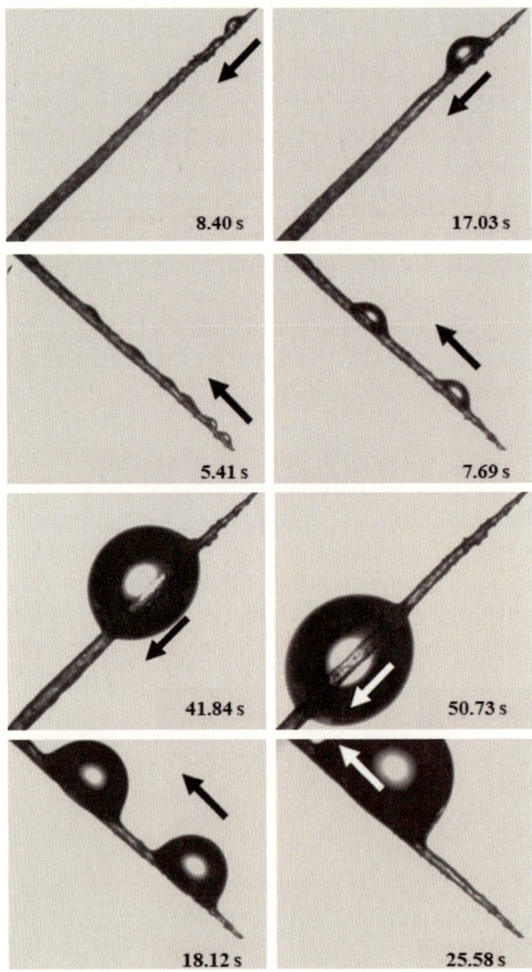

Spine Anatomy

Opposite Page:
Top Image-
Magnification of barb highlighting tip, groves and the base

Middle Image-
Magnification of cactus spin barb showing apex angle

Lower Image-
Magnification of spine multi-level groves

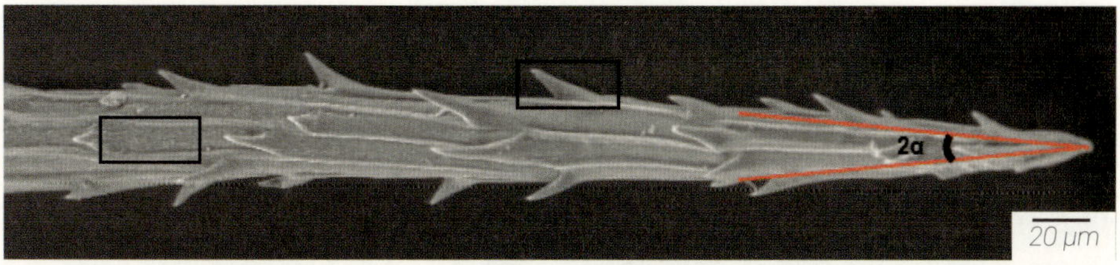
2α
20 μm

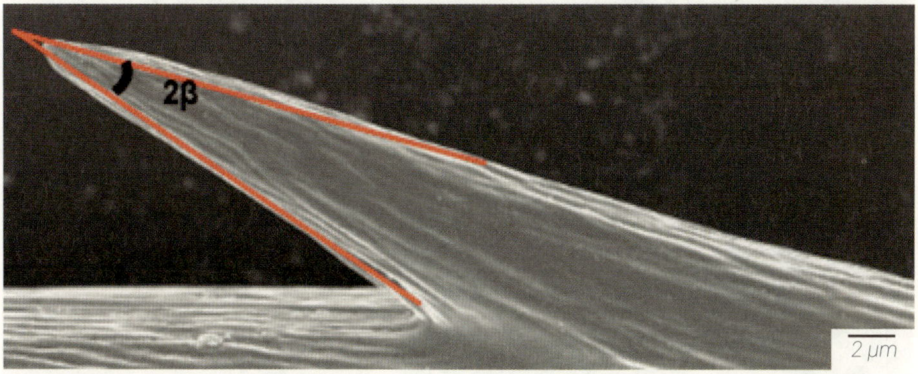
2β
2 μm

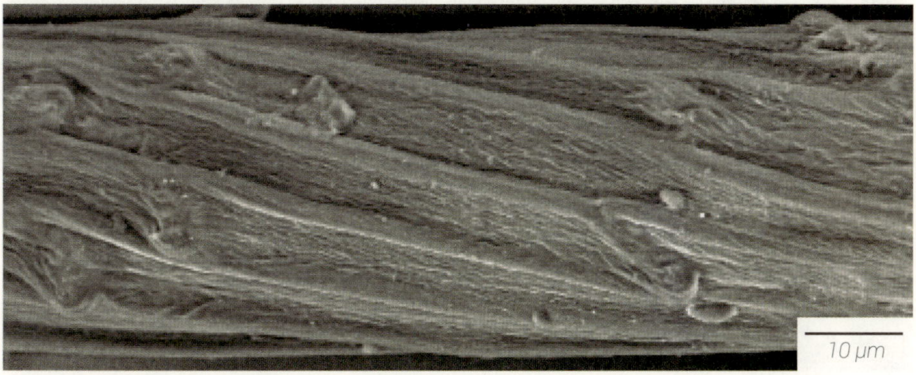
10 μm

Compound Eyes

Mosquito Vision

Mosquitoes have the ability to locate various resources such as mates, hosts, and resting sites in wet and dimly lit habitats. They are able to accomplish this thanks to their excellent vision. This feature is attributed to the structure of their compound eyes which have ideal superhydrophobic properties that provide an adequate protective mechanism for maintaining clear vision in humid habitats.

Their eyes are a combination of hexagonal protrusions at the nanoscale that form the individual lenses of ommatidia (each of the optical units that make up a compound eye), which play a crucial role in creating the ideal superhydrophobicity for preventing microscale fog drops (moisture) from condensing on the eye surface. At the microscale, the hexagonally close-packed (hcp) protrusions prevent water droplets from collecting in the voids between the ommatidia.

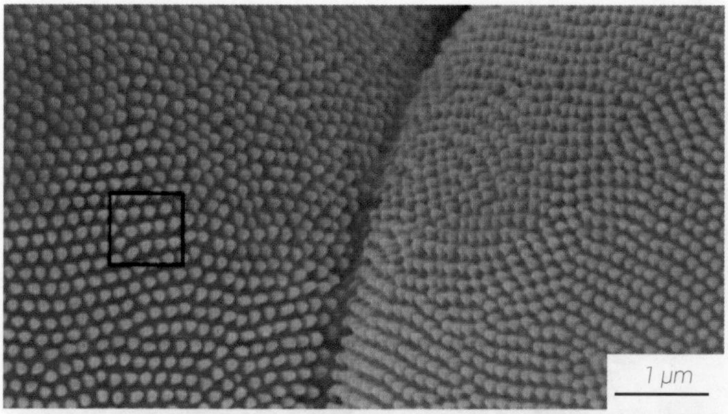

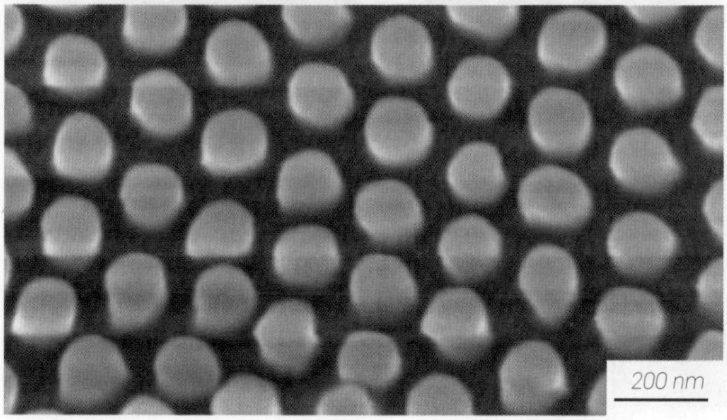

Mosquito

Opposite Page:
Photomicrograph of mosquito.

Current Page:
Mosquito eye hexagonal protrusions enlarged in microscan images.

Research by Lei Jiang.

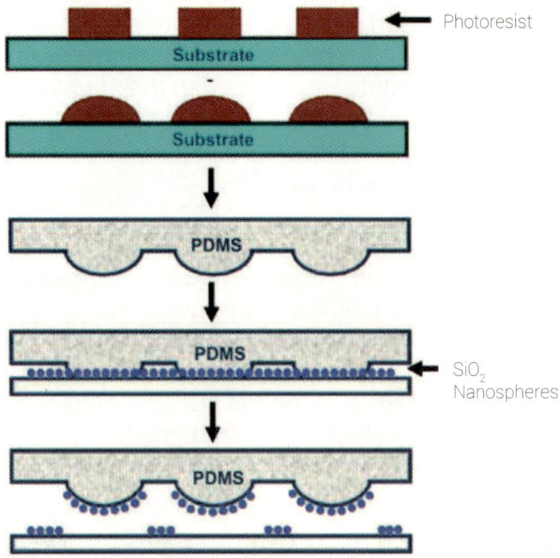
Photoresist
Substrate
Substrate
PDMS
PDMS
SiO2
Nanospheres
PDMS

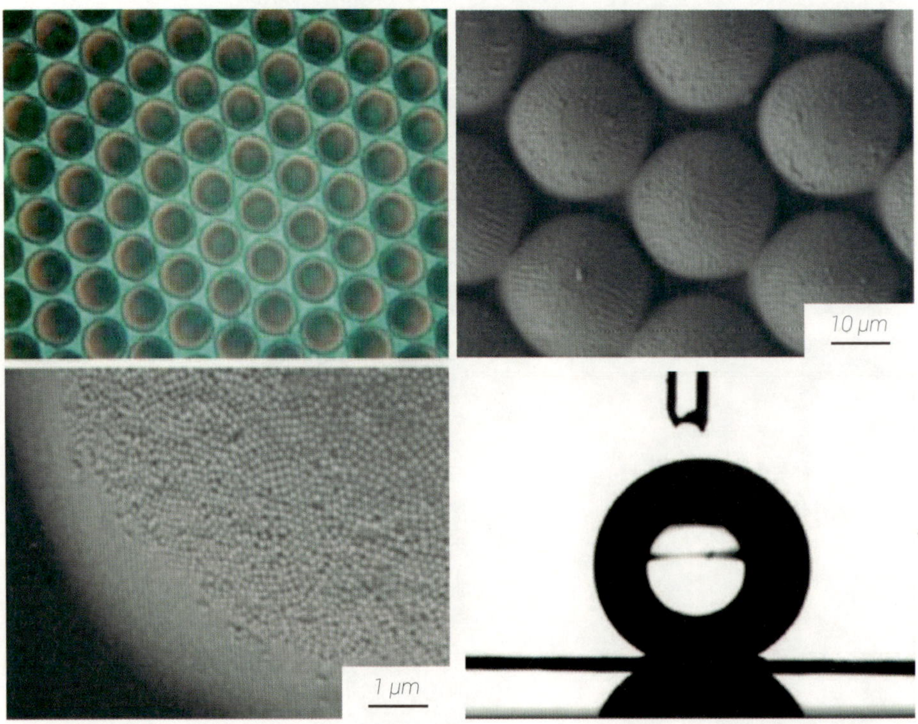
10 μm
1 μm

We developed an artificial surface material with similar properties of superhydrophobicity. We used photolithography to fabricate the hexagonal packed circular protrusions through a series of photoresist posts on a glass slide. Photoresists are light-sensitive materials used to form a pattern coating on a surface.

We then patterned a monolayer of colloidal crystals on the surface of PDMS (silicone polymer) hemispheres to mimic the nanostructures by using a soft-lithography technique. A silicon stamp was then patterned with micro-hemisphere arrays and brought into contact with a monolayer of silica nanospheres on the silicon substrate. After the silicon stamp cooled, it was carefully peeled from the substrate, and we obtained an artificial composite material.

We observed that the silicon hemispheres had a uniform diameter of ca. 22 μm (ca= critical angle). The average value of the static water ca. and the tilted angle was ca. 155° and ca. 15°, respectively, which indicated that the microhemisphere and nanosphere structures could significantly enhance the water-resistant property of the surface.

We believe that these findings could provide an insight for designing superhydrophobic materials which may be developed into multifunctional coatings with anti-fogging, anti corrosive, easy-cleaning, and anti-reflective properties.

Manufactured Surface

Opposite Page:
Top Images
Diagrams showing the process of fabrication for the artificial compound- eye material.

Lower Images
An optical photograph of an artificial compound eye, composed of a large number of uniform PDMS (silicon) hemispheres, shown at micro and nanoscale

Photograph showing a spherical water droplet on the artificial compound-eye microstructure.

Spatulae

Gecko

Geckos are lizards that have successfully adapted to very diverse habitats throughout the world thanks to their unique physical features. One of these skills is their ability to climb vertically on smooth surfaces at great speed, which is attributed to the hierarchical structures found in their feet. The gecko foot consists of well-aligned microscopic keratinous hairs called setae which are split into hundreds of smaller nanoscale ends called spatulae. Contact between the gecko spatulae and an opposing solid surface generates the Van der Waals forces that are sufficient to allow the gecko to climb vertical walls or across ceilings. In addition to this remarkable adhesion property, the gecko feet also exhibit superhydrophobicity and self-cleaning attributes.

Our lab's research has developed a super-hydrophobic polystyrene (PS) layer, which is composed of more than 6,000,000 aligned nanotubes per square millimeter, similar to the feet of a gecko. We believe that these findings could provide an inspiration for designing superhydrophobic materials in the future, which may be also developed into multi-functional coatings.

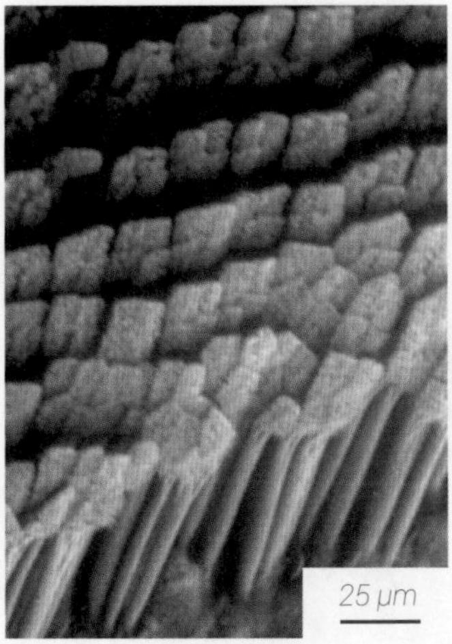

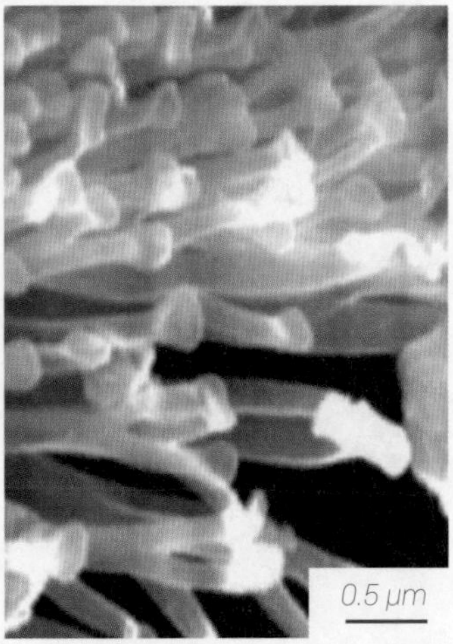

Gecko Spatulae

Opposite Page: Image of Gecko leg and foot

Current Page: Scanned images increasing in scale showing the tiny keratinous hairs called setae on geckos' feet which help in maximizing contact with surfaces.

Research by Lei Jiang.

Donghua University

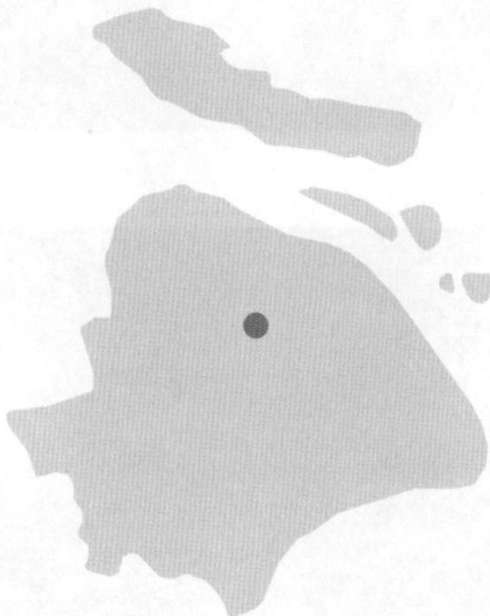

The Dynamic Technology Lab for Bionic Materials and Structures (DT-Lab) was created in 2015 by Professor Dengteng Ge, as part of the Institute of Functional Materials at Donghua University. The lab has received grants from the National Natural Science Foundation of China (NNSFC), and by the Natural Science Foundation and the Science and Technology Innovation Program of Shanghai. Currently, the team consists of 12 people, with ongoing collaborations with universities in China and the USA. The lab's primary focus of study includes artificial microstructures, their deformation control, the manufacturing of colloids, and polymer-based functional composites.

Water Crystals

Opposite Page:
Ice formed packing pattern

Carystoides Escalantei
Butterfly

Top Image:
Female *Carystoides escalantei*.
Photo taken at vertical direction with flash light.

Middle Image:
Male *Carystoides escalantei*.
Photo taken at vertical direction with flash light.

Lower Image:
Male *Carystoides escalantei*.
Photo taken at low angle with flash light.

Extraordinary Whiteness

Whiteness is frequently appears on the wings, legs, and bodies of many species of moths and butterflies. While "white" may often be simply one more color among many within a larger complex pattern, there are times when the whiteness itself appears to be a key signal.

Natural whiteness is often attributed to the scattering or diffusion of light from (sub)micron-sized textures including ribs, ridges, and pores. Species, such as *Pieris brassicae* and *Delias nigrina,* have randomly packed beads in between ribs and cross-ribs, which enhance their appearance of whiteness. Metallic to silvery whiteness occurs on some butterfly wings as the result of mixing structure colors for the purpose of communication.

Here, we investigate the nano and microstructures of white spots on the wings of *Carystoides escalantei*, a dusk-active and shade-inhabiting Costa Rican rain forest butterfly (*Hesperiidae*).

On both males and females, two types of whiteness occur: angle-dependent and angle-independent. Interestingly enough, some spots on the male wings are absent from the female wings.

Carystoides Escalantei

Male *Carystoides escalantei* with white on antennae, middle body region and tips of wings.

Research by Dengteng Ge.

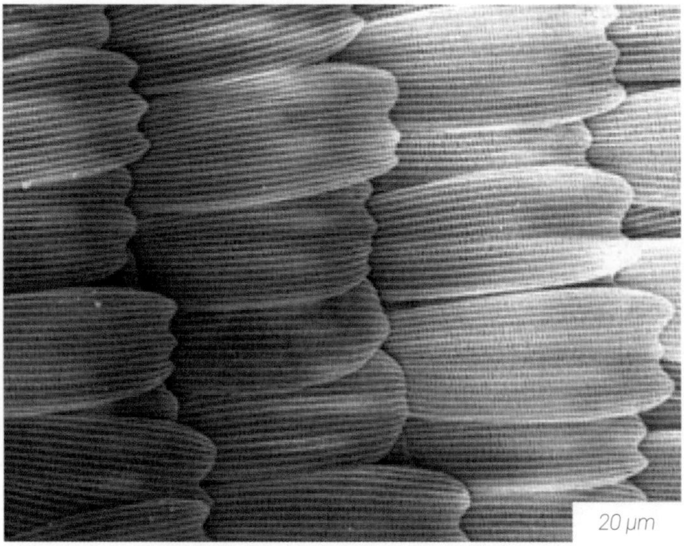

Angle Independent Structure

SEM Photos of the white spots at the antenna.

The angle-independent areas at the antenna and wing tips have white spots, which differ in the microstructure, orientation, and associated properties of their scales. The angle-dependent whiteness can be bright or dull depending on the observation directions.

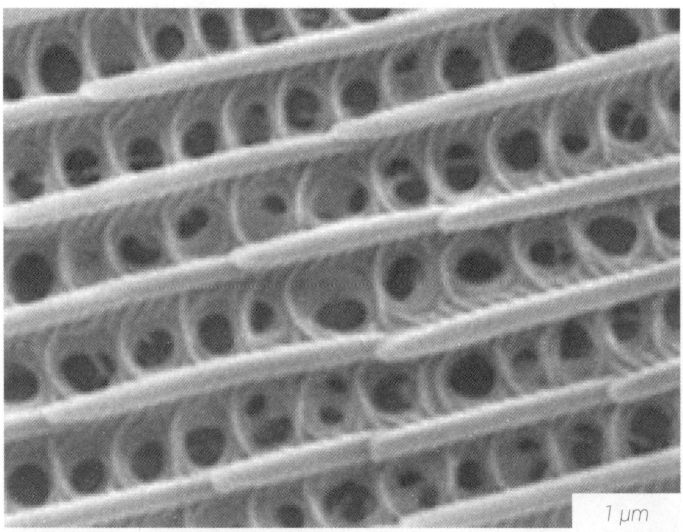

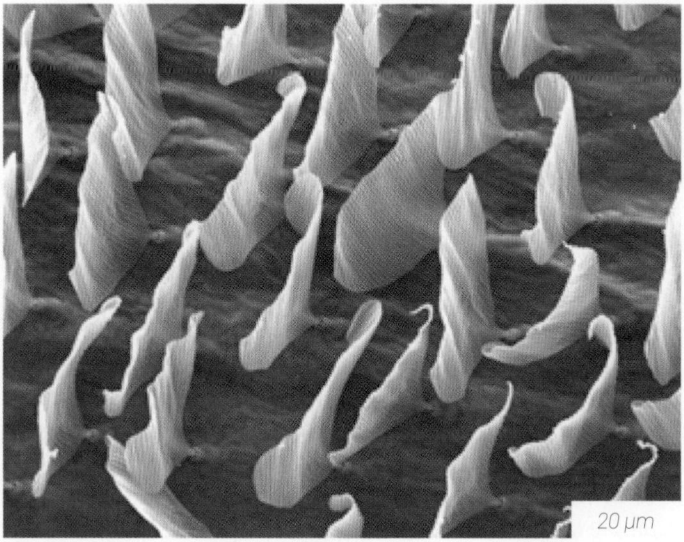

Angle Dependent Structure

SEM Photos of the white spots in the middle of front wing

Angle-dependent areas are the middle region of the wings.
The angle-dependent scales show enhanced retro-reflection.

Angle-dependent areas in the middle region of the wings and angle-independent areas on the antenna and wing tips have white spots that differ in their microstructure, orientation, and associated properties of their scales. The angle-dependent whiteness is bright or dull depending on the observation direction. The angle-dependent scales also show enhanced retro-reflection. We speculate that the biological functions and evolution of *Carystoides* spot patterns, scale structures, and their varying whiteness are adaptations to butterfly's low light habitat.

Typically, whiteness is the result of scattered random microstructures accompanied by no pigment or dye beneath the scales. We observed that when the structures of the scales decrease to submicron size, the scattering is also decreased, and whiteness becomes a result of mixing multiple colors with angle-dependent spots. The bright vs. dull aspect of whiteness depends on whether the scales are more or less vertically oriented in the direction of the wing base or the wing tip.

Structural whiteness is technologically vital in systems ranging from power-efficient computer displays to sensors, to energy efficient buildings, windows, and vehicles. The patterns and dependency on angles found in this study could help benefit these technologies and advance concepts of structural color.

Scales Form

Opposite Page: Bright vs dull features of whiteness depend on whether more or less vertically oriented scales are observed from the direction of the wing base or the wing tip.

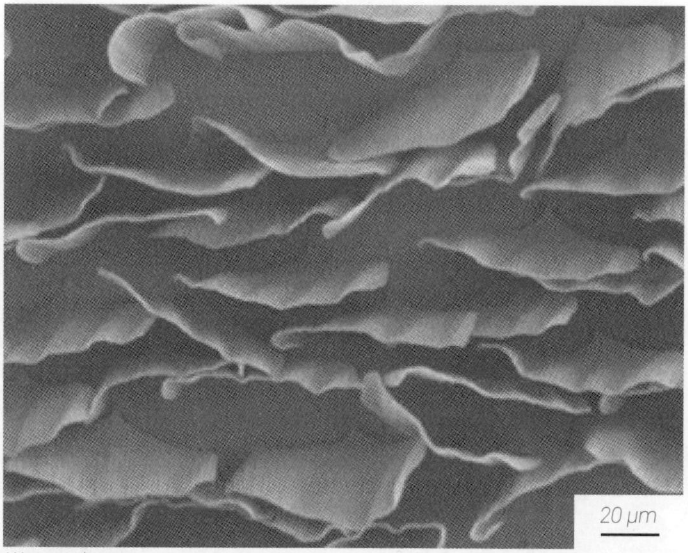

Wavy scales

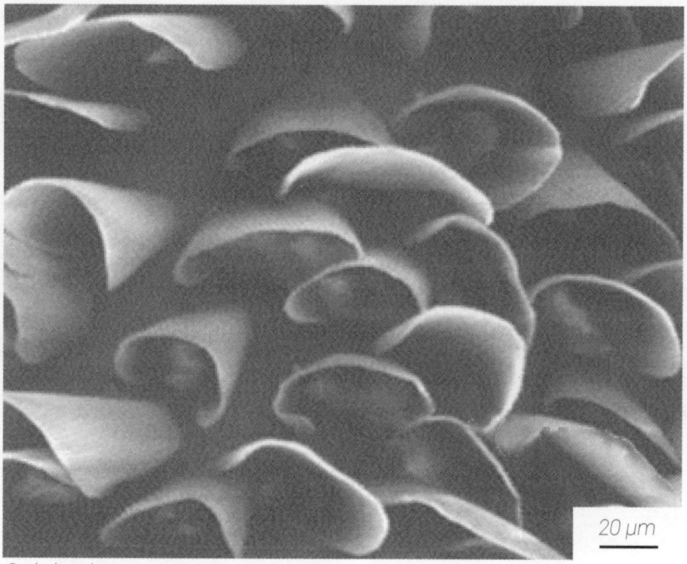

Curled scales

Scales

Image of wing section showing scales and wing membrane with ventrals for airflow.

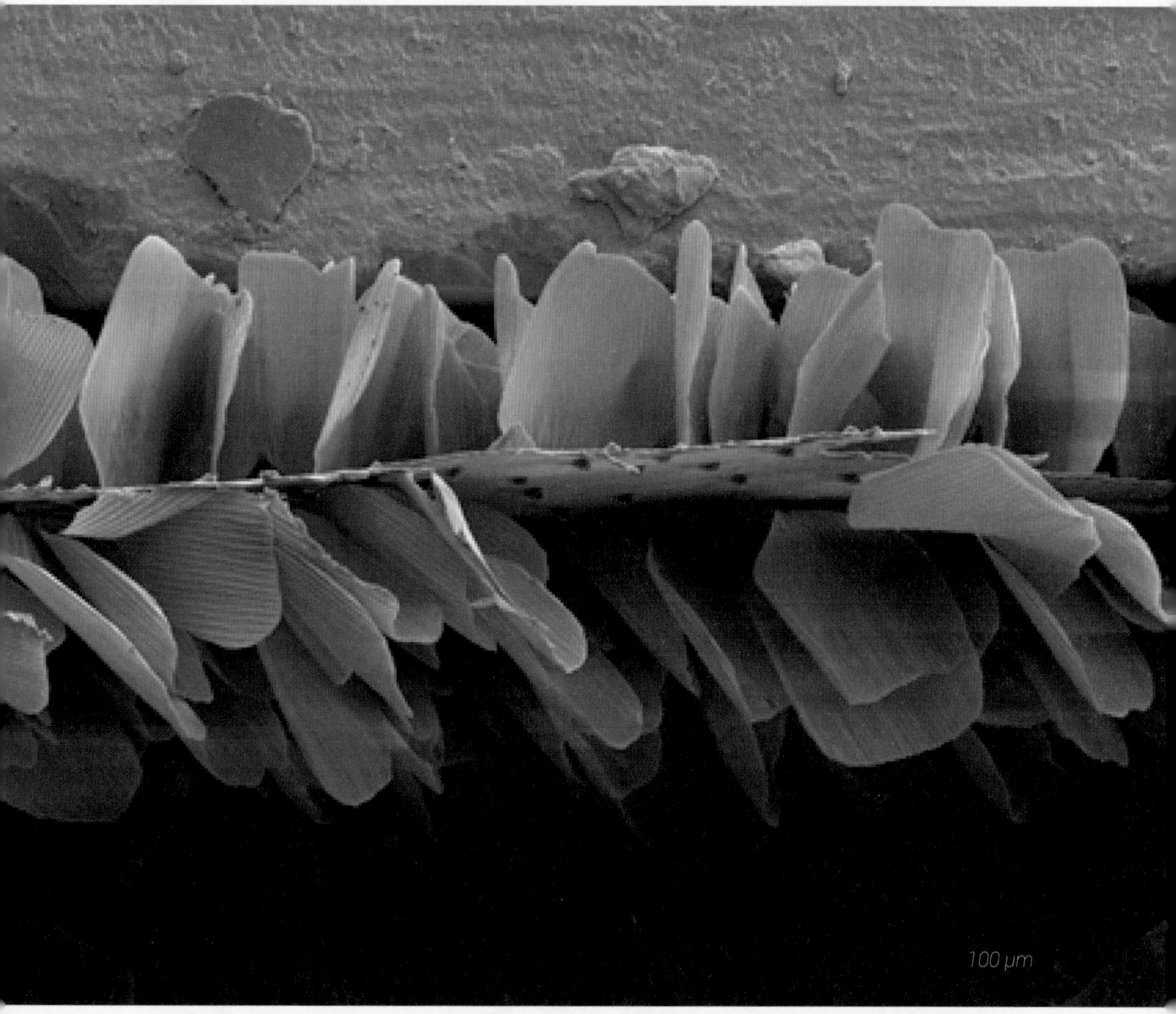
100 μm

Pigmented Skin

Cephalopods & Dynamic Structures

Cephalopods (octopuses, squid, and cuttlefish) can switch between transparent and pigmented states by expanding and contracting their cells. This is known as optical modulation and is a common practice to control light transmission macroscopically, similar to the opening and closing of curtains or blinds.

To explore this concept we are developing composite films consisting of a thin layer of arrayed silica nanoparticles (NPs) embedded in bulk elastomeric PDMS. This type of exploration works at the microscale, by the tuning of the optical properties through mechanical stretching and compressing of polymer films.

In the materials initial state, it is highly transparent. Upon mechanically stretching it, the transmittance is dramatically reduced to 30% and displays a structural color. We found that the color could be tuned by the silica NP size. Unlike the silica NPs/PDMS films prepared from highly ordered silica colloidal crystals, the reflective color did not change with increased strain. The switch between transparency and colored states could be reversibly cycled at least 1,000 times without losing the film's structural and optical integrity.

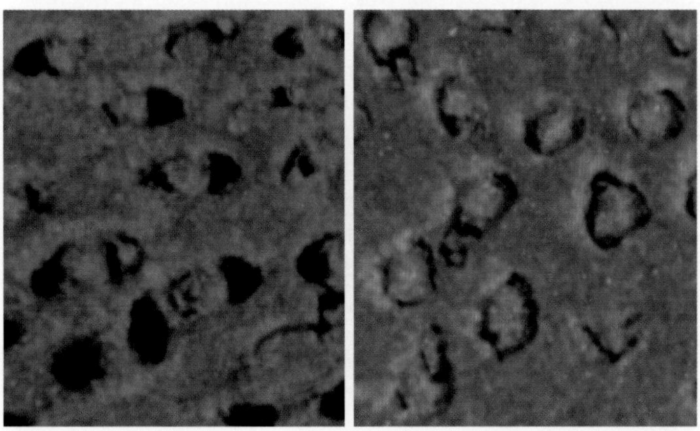

Transparent & Pigmented States

Opposite Page
Transparent switching in Cephalopod's skin is standard in many octopus, squid and cuttlefish.

Current Page:
Images showing the material testing of reversible nanovoids in Silica NPs/PDMS.

Research by Dengteng Ge.

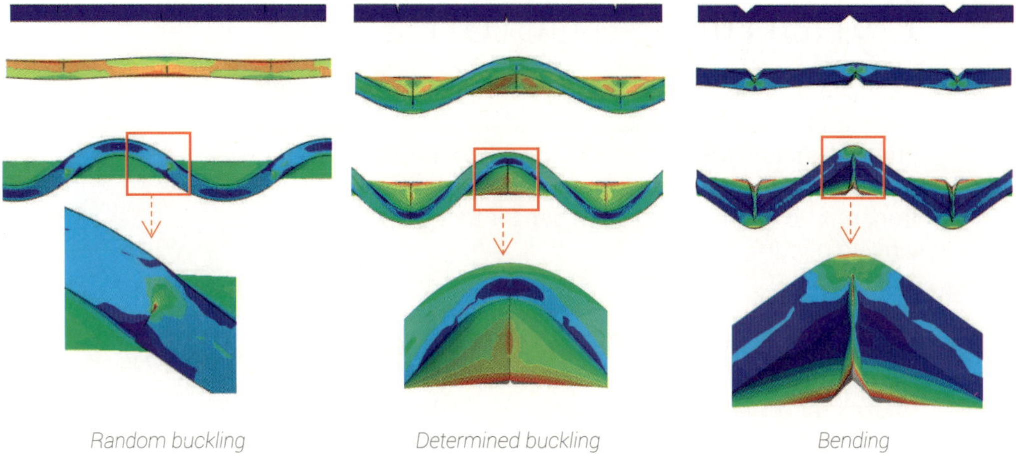

Another common practice in dynamic material research is how to have a material respond to external stimuli. The question begins with how to reconfigure metamaterials (a material engineered with properties not found in nature) that can allow for a dynamic change in the structural patterns. Such materials can offer unprecedented capabilities to steer the flow of electricity or light.

Here, we have created a variety of reconfigurable one-dimensional (1-D) soft and adaptive metamaterials through a cut guided buckling in an array of discrete plates. Through engineering the cut geometry and the relative placement, we can locally and reversibly activate and lock the buckling of the soft metamaterials, thus selectively steering the electronic and optical pathways. Our soft metamaterials are "pluripotent", i.e. the pathways can be written, visualized, erased, and rewritten into a completely new one by merely varying the local stimulus.

The research has shown that material pigmentation control, and external stimuli informed by logic of Cephalopods could be useful for the future developments of applications in mechanical smart windows, anti-counterfeiting and projection.

Adaptive Metamaterials

Current Page: Diagrams showing reconfigurable 1-D soft plates their adaptive metamaterial deformation.

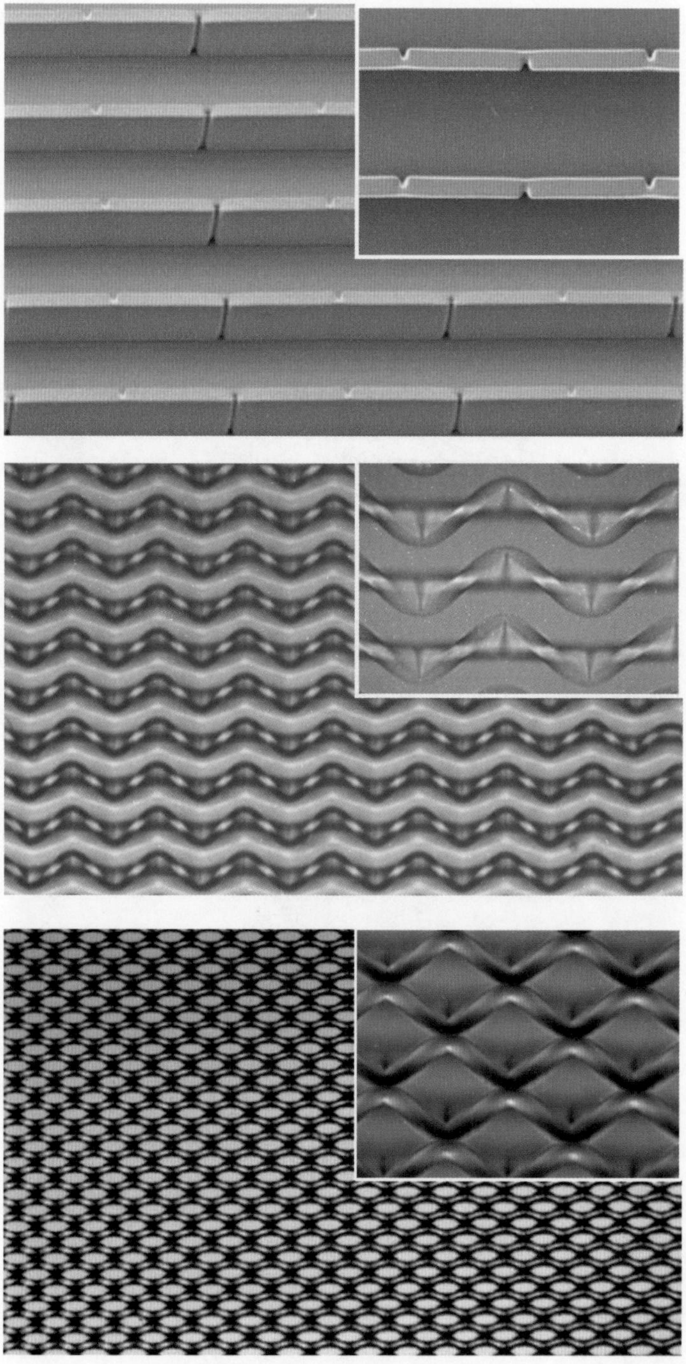

1-D Plates

Current Page:
1-D plates depicting the buckling range induced after stimuli.

Turning Heads

Tongji University

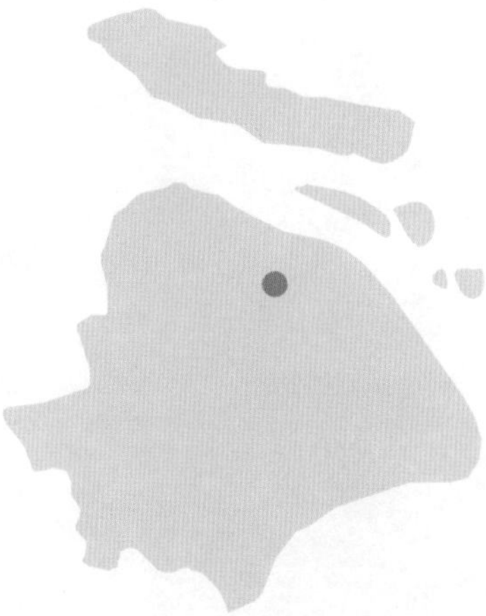

The Biomimetic Design Lab (BiDL) was established in 2012 by Pius Leuba and Xiaocun Zhu, as part of the Design and Innovation College (D&I) at Tongji University in Shanghai. The BiDL lab was established as one of the first centers to integrate biomimicry and sustainability with design and innovation at a Chinese university. The lab focuses on environmental design explored through biomimicry, fostering holistic design approaches to reconnect humankind and nature. The lab has ongoing collaborations with national and international institutions, including IDEA Design Lab, GIGA (Green Ideas, Green Action) and Biomimicry 3.8. The lab was instrumental in the creation of the Biomimicry China Network(BCN), providing a national platform for Chinese professionals interested in Biomimicry. In 2015 Mary Polites joined BiDL as co-director and biomimetic thesis advisor. Her research focuses on environmental responsive material systems and the following projects are examples of this agenda through student thesis work.

Heliotropism

Image:
A field of blue buttercups tilt to face the sun

Research led by Mary Polites.

Heliotropism

The construction industry represents a high amount of energy consumption due to its systems such as heating, cooling, ventilation, as well as lighting and other household appliances. In response to this, energy savings strategies are increasingly becoming a focus to reduce consumption by increasing efficiency.

This research explores prototypes using locally accessible materials, to improve thermal comfort while reducing the reliance on systems that require energy to alter environments.

The movement of plants informed the exploration of this work. In general, plants are in a fixed position, therefore to increase the chance of exposure to the sun they have solar tracking mechanisms. This movement includes phototropism and heliotropism. Phototropism is a growth-mediated movement in response to unilateral light. Heliotropism is a dynamic and oscillatory form of plant movement, which is, distinguished from phototropism, as a reversible, rapid and not involved in the growth periods of plants. Both allow plant organs to remain close to optimize conditions for growing (e.g. energy obtaining an optimal temperature).

Early on the projects intended to have mechanisms which would have the ability to articulate movement in response to a stimuli. This movement needed to alter the thermal or environmental conditions of a space.

The following work represents two projects which address this goal. The first "Microclimate Regulation System Based on Plants Adaptability" by Cunyi Liu and the second, "Responsive Kinetic Structures Based on the Tropic and Kinetic Movement of Plants" by Jiarui Tan.

Panel Prototype

Image of modular panel system final model.

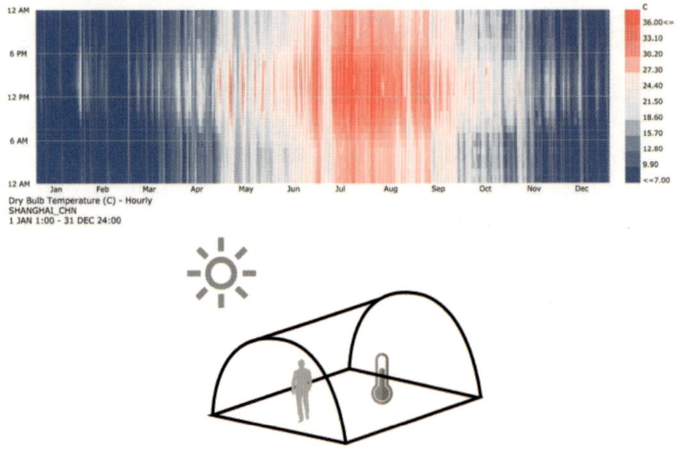

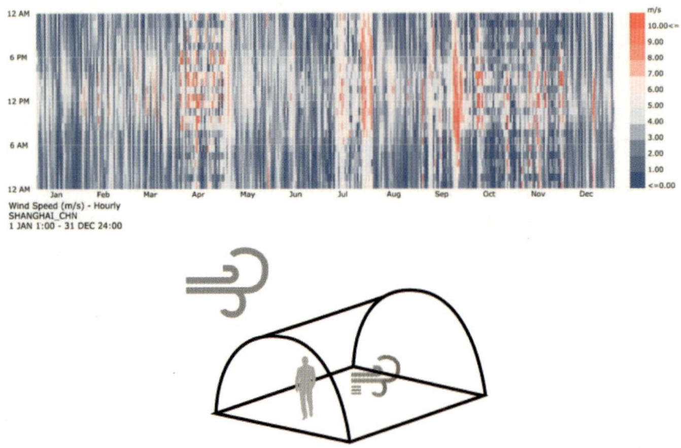

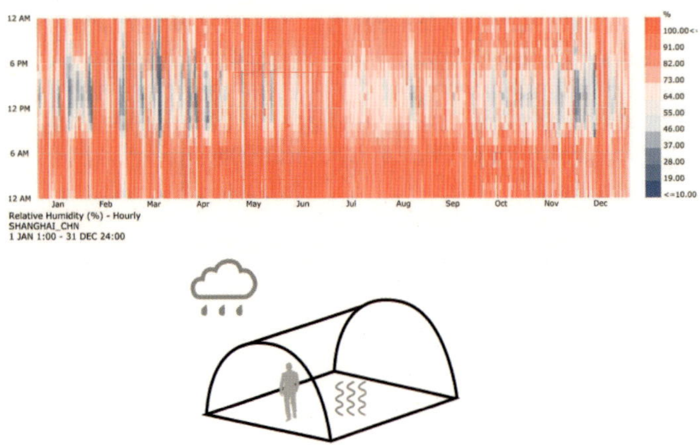

Shanghai Climate

Top Image:
Annual Temperature
Shanghai China

Middle Image:
Annual Wind Speed
Shanghai China

Lower Image:
Annual Relative
Humidity Shanghai
China

Microclimate Regulation System Based on Plants Adaptability

Solution-Driven Design

For this project, we explored biomimicry through a method called solution-driven design. This method is useful when articulating new concepts without the pressure of sourcing an immediate application.

As part of the background research, we looked at the weather data to develop guidelines for the design. The conclusion from this work was that the hot humid summers of Shanghai were more intense than the damp winters due to consecutive days, and extreme temperatures. Additionally noted was the direction of the wind, which shifted during the summer to winter months.

During our observations of plants, the movements of leaves were one aspect that lent itself to many possibilities. Leaves can move in ways that allow for energy absorption positions as well as shading arrangements. Considering these benefits, we explored the movements and rotational range of leaves and then categorized these into a series of angles for further study.

This range of movement was modeled through prototyping materials. The first prototype explored a single module which could later be compounded into a system. Only after exploring the material abilities was wind defined as external stimuli. The modules could open and close depending on the overall placement of the system in relation to the time of the year. Summer—remain open and guide wind through for cooling; Winter—close and block exposure. This application could be beneficial for exterior urban spaces, where conditioned exterior space is necessary.

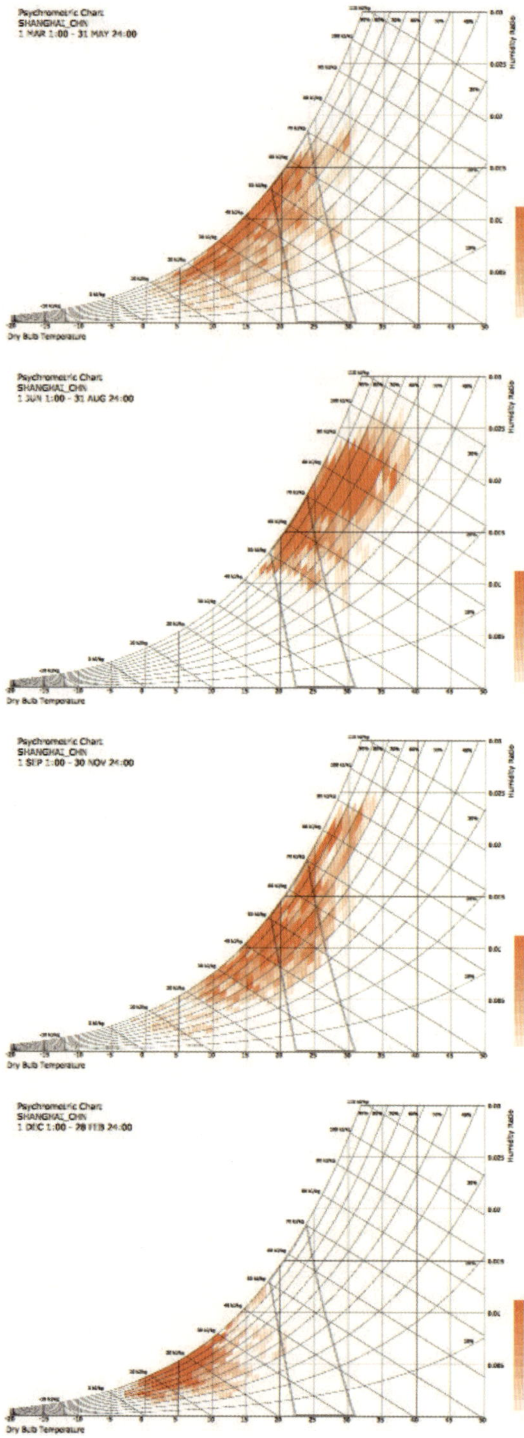

Psychrometric Chart

Seasonal charts showing ranges in comfort in the highlighted areas

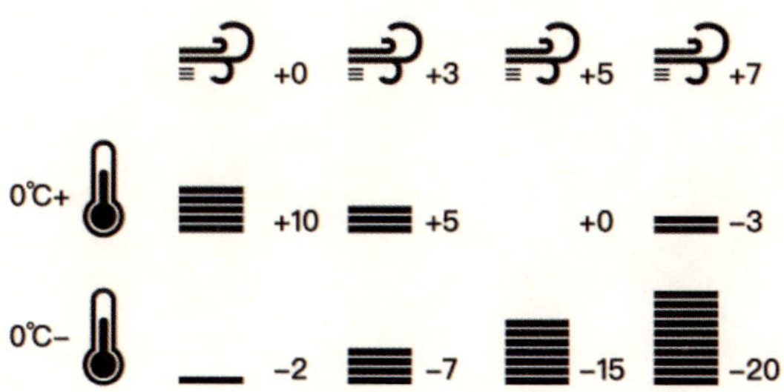

Weather Concepts

Human comfort is determined through body temperature, humidity, precipitation, solar exposure, and wind speed.

Top Image:
Relationship between temperature and environmental factors

Middle Image:
Relationship between humidity and environmental factors.

Lower Image:
Relationship between wind speed and body temperature

Site Conditions

Wind Rose Diagrams

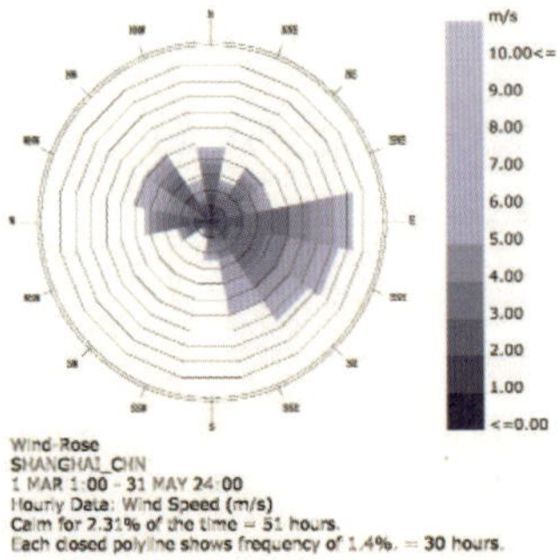

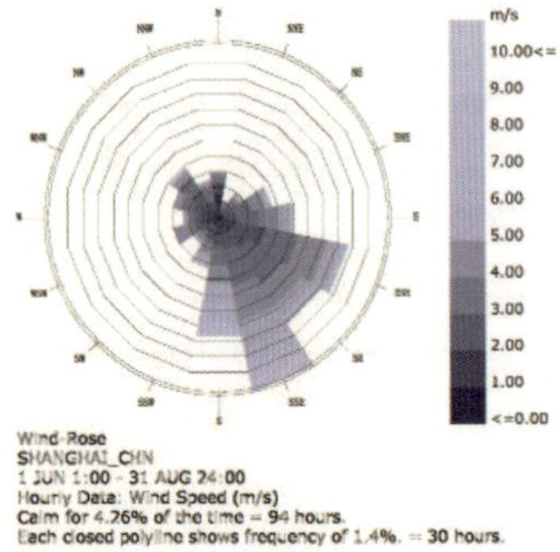

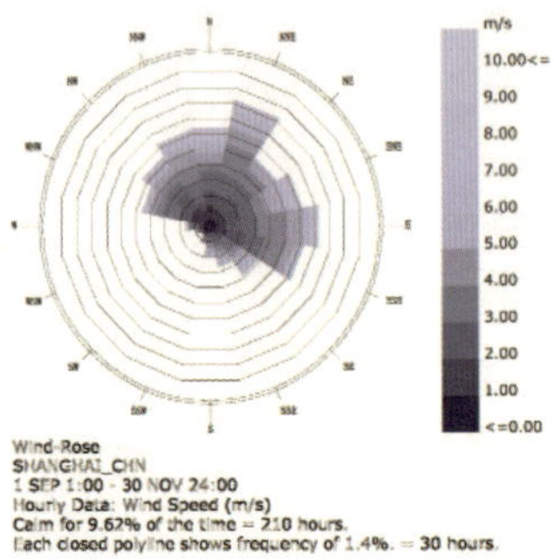

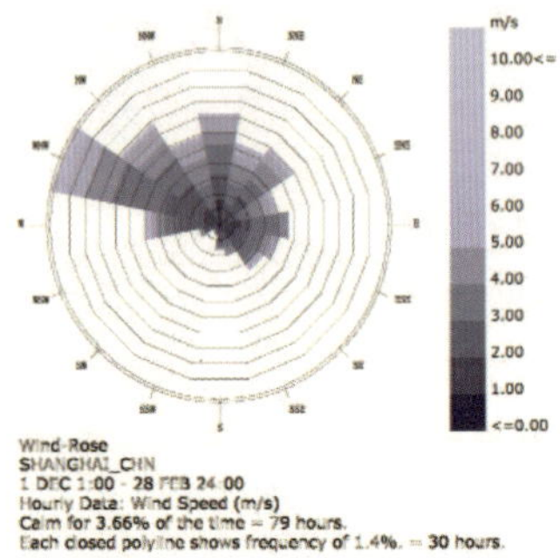

Site Conditions

Lower Images: Massing surface areas of protection throughout the year

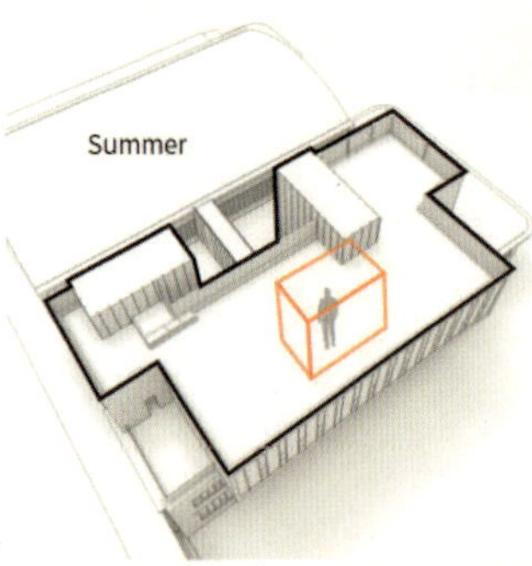

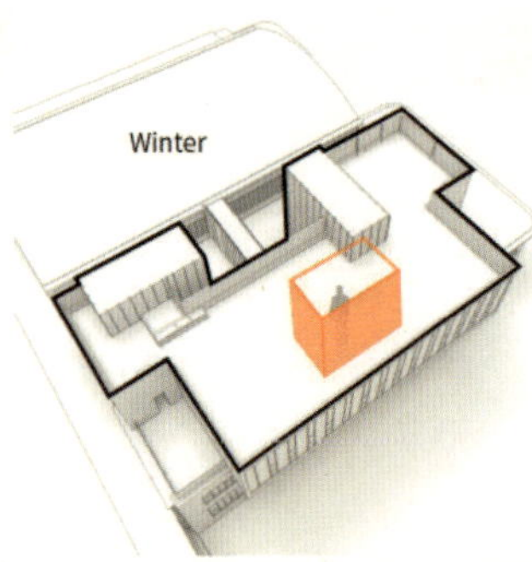

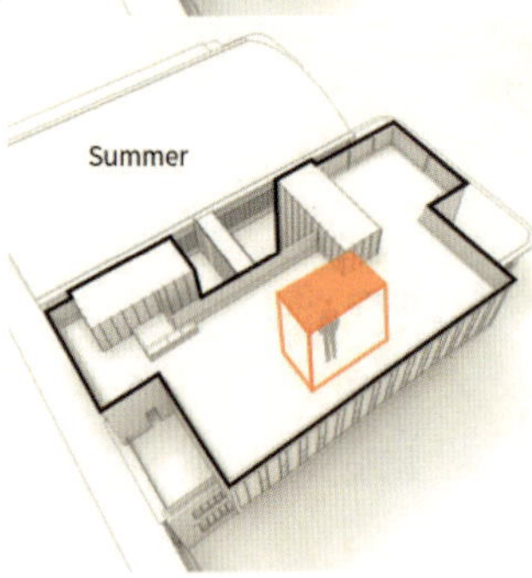

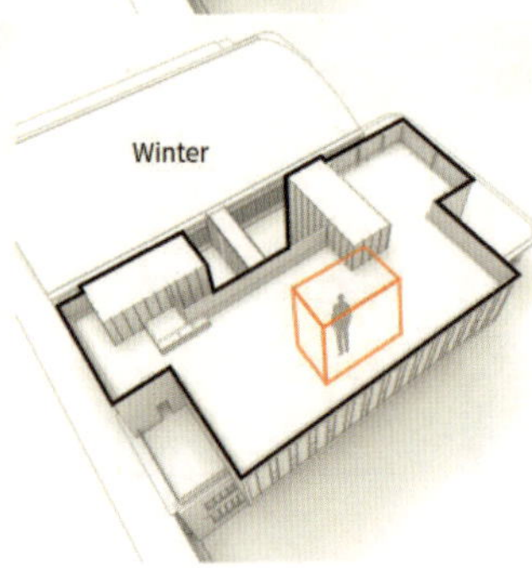

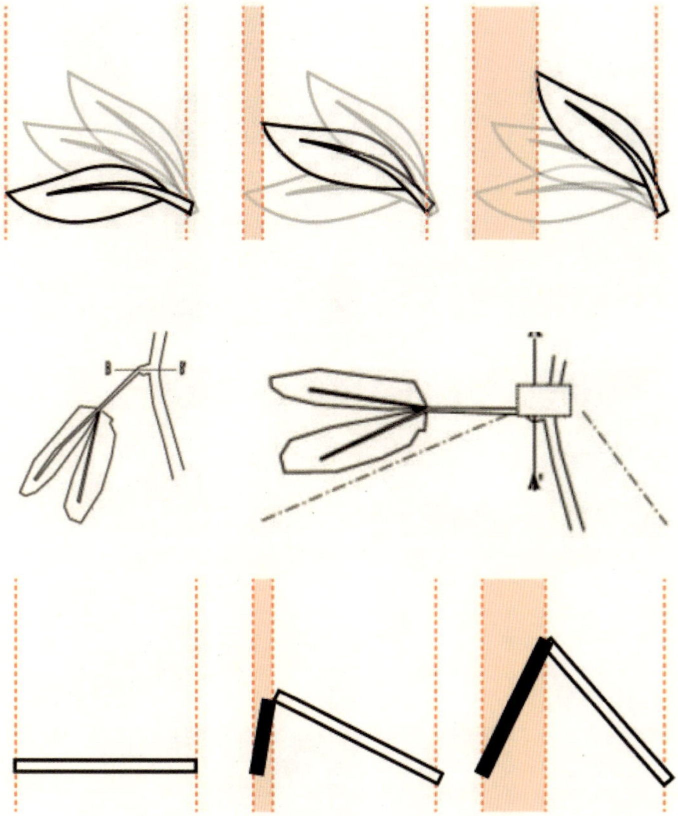

Movements Diagrams

Leaf movement concepts translated into shading systems

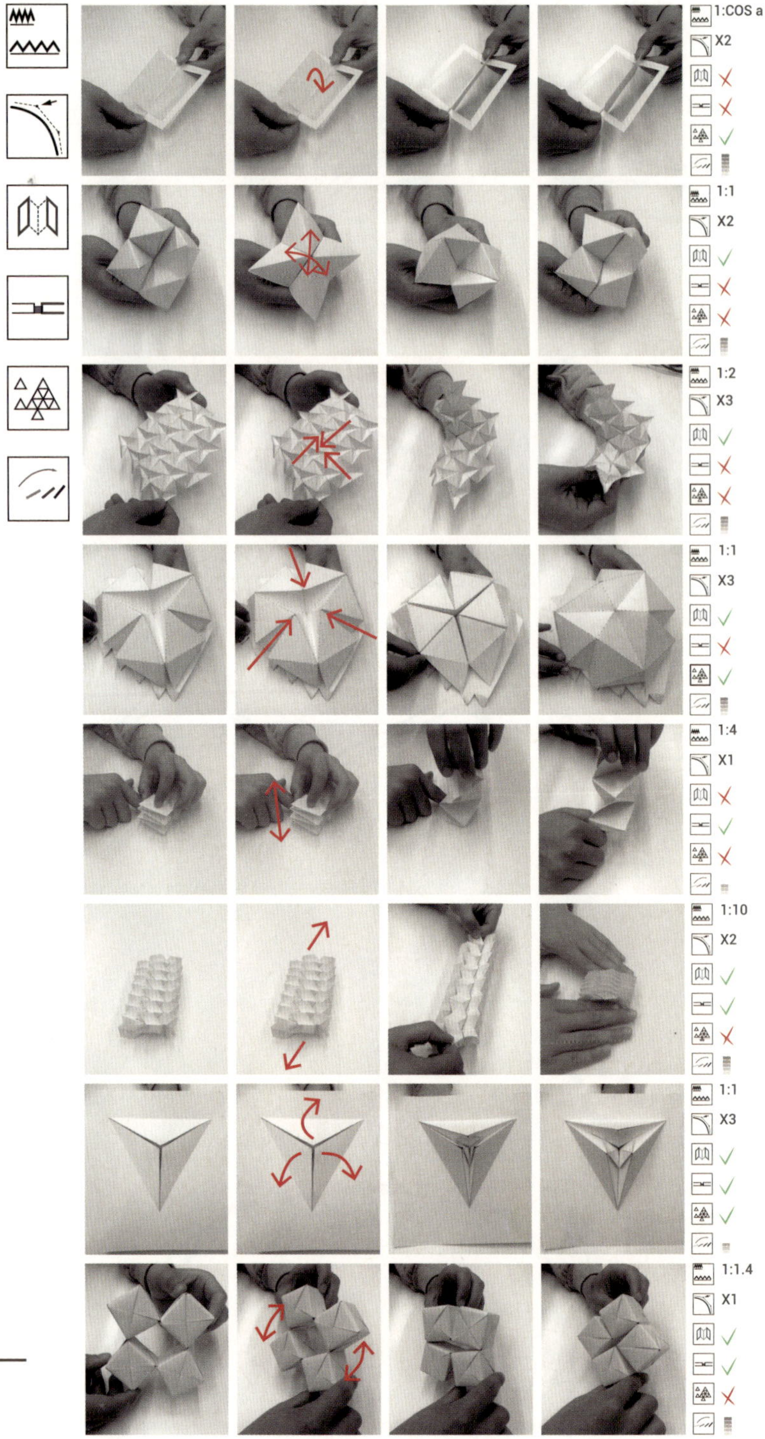

Model Studies

Paper Folding
Movement Studies

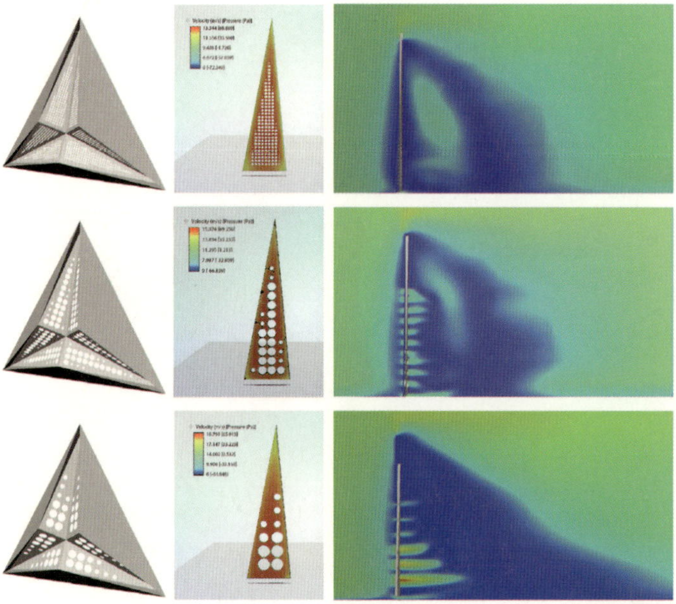

Module Studies

Wind studies through opening patterns in modules

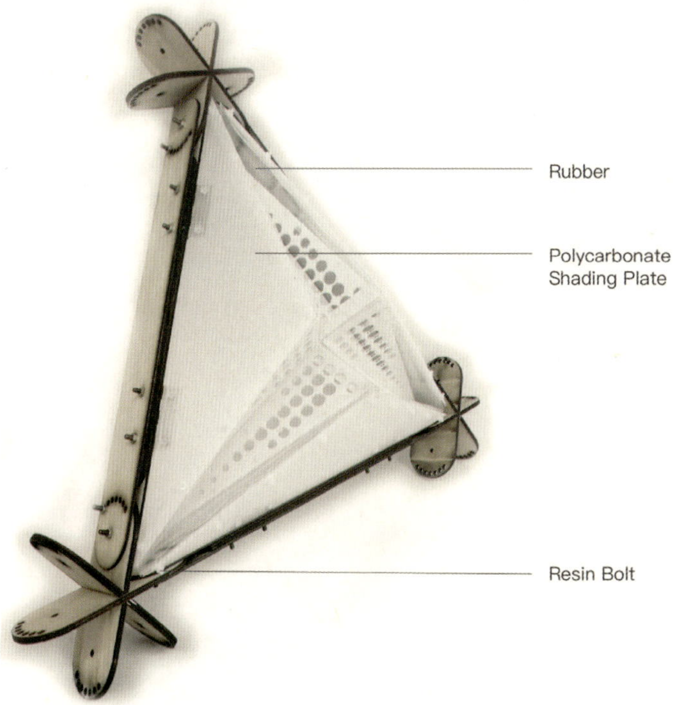

Physical Model

Physical Model with Perforation

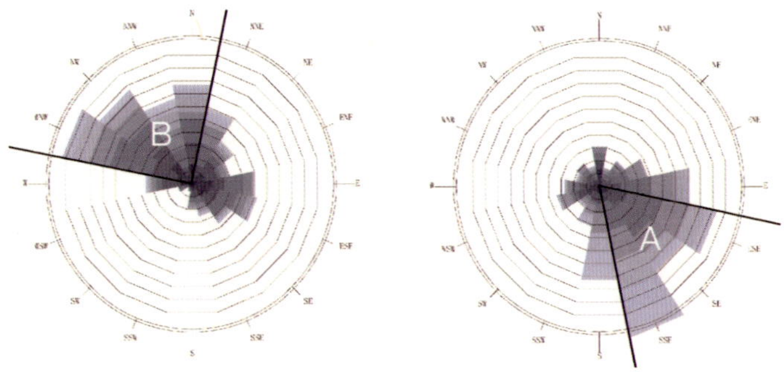

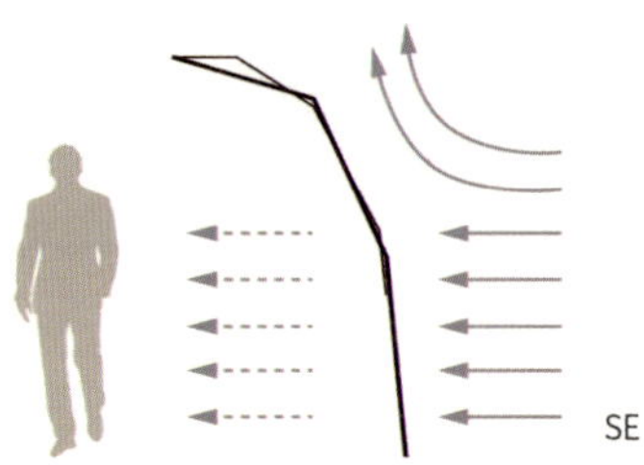

SUMMER SECTION A

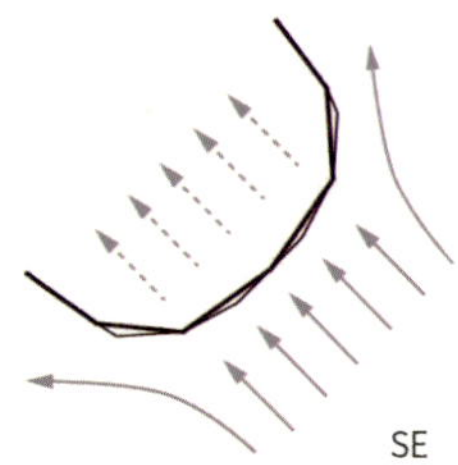

SUMMER PLAN A

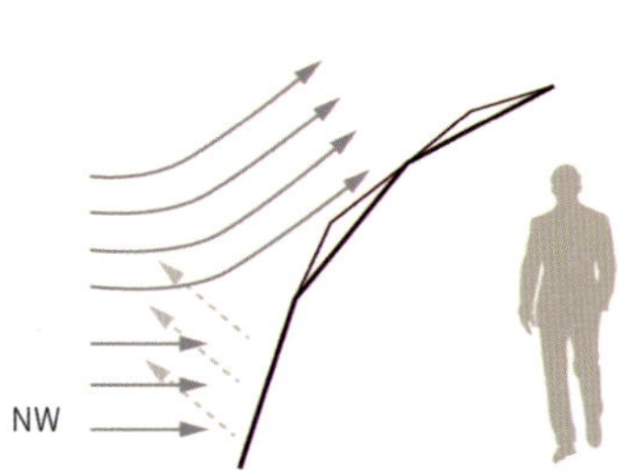

WINTER SECTION B

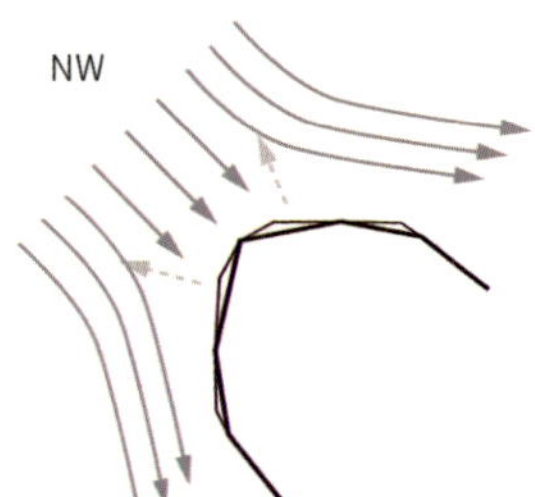

WINTER PLAN B

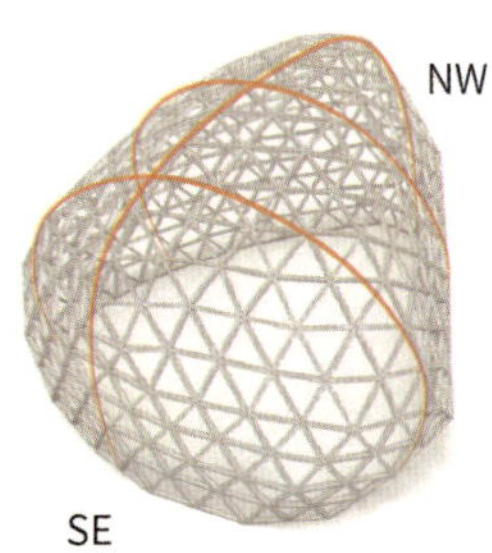

Site Conditions & Design

Wind direction aligned with module direction to connect to space inside

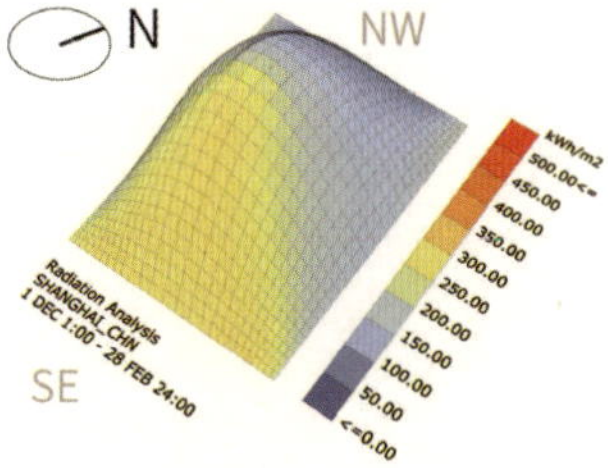

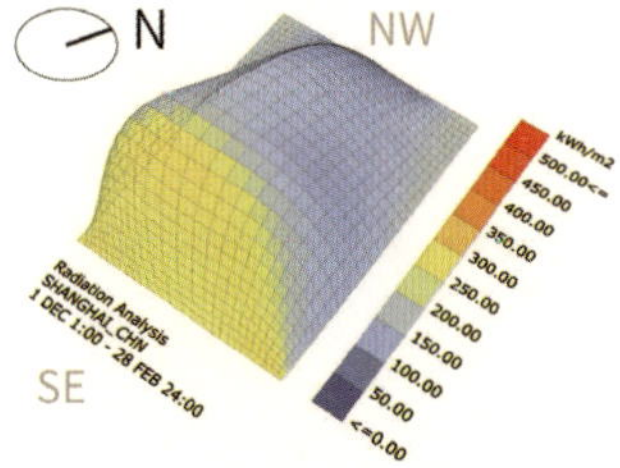

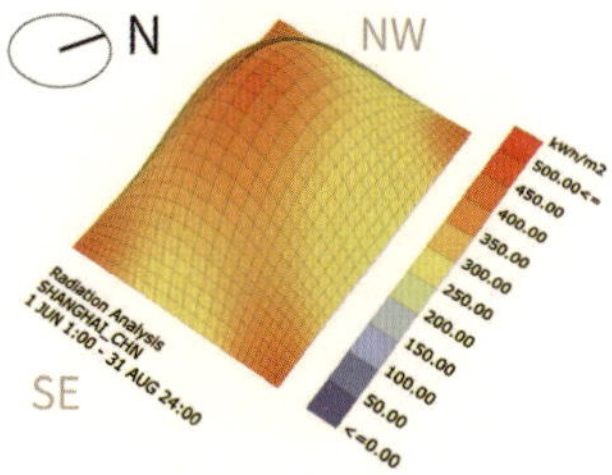

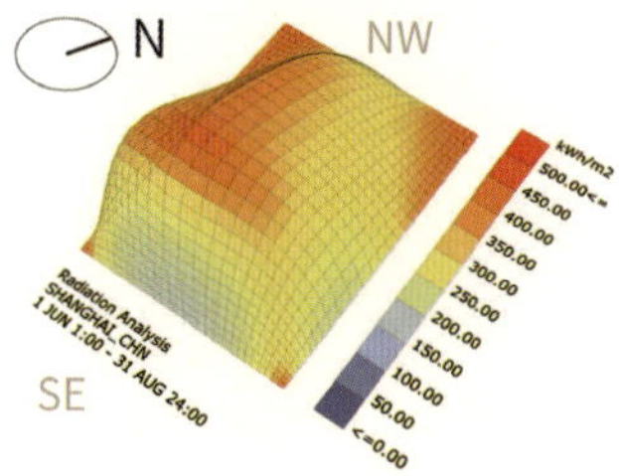

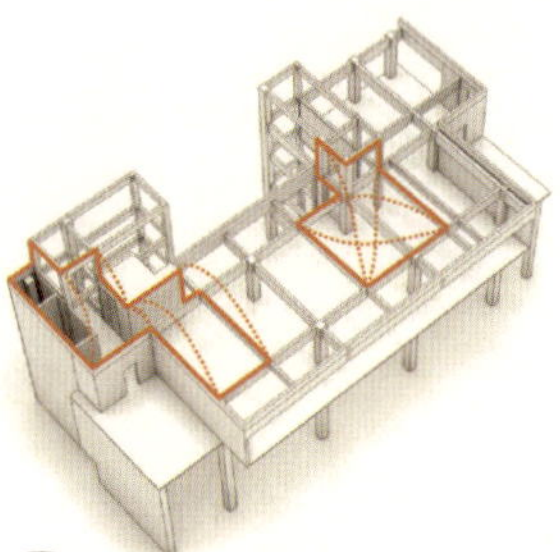

Beam & Column & Stair Case

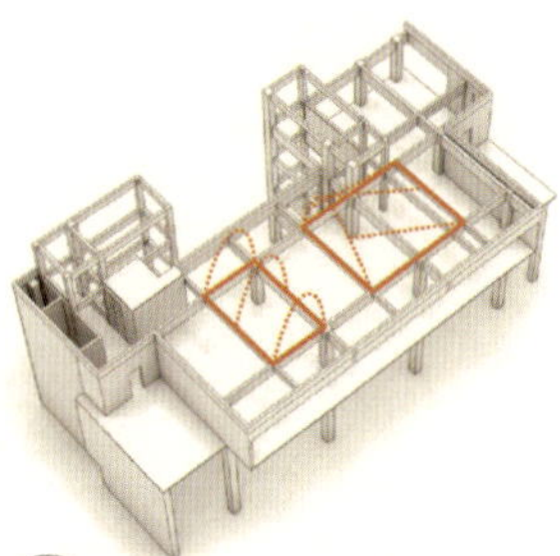

Beam & Column

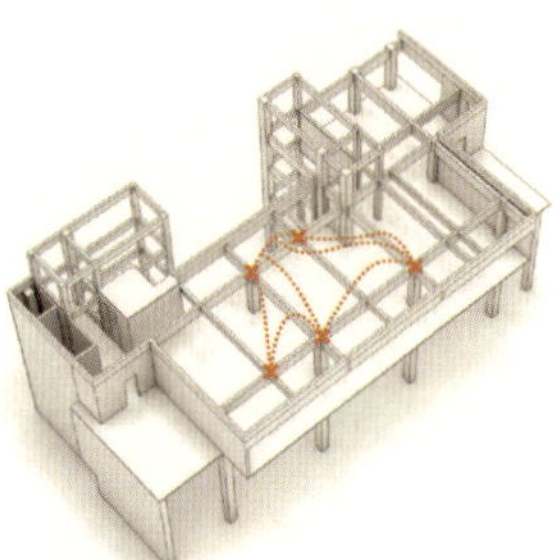

Column

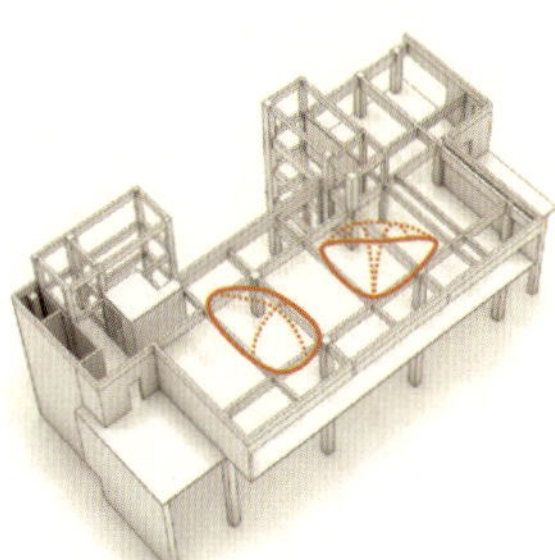

Beam

Site Conditions

Solar exposure aligned with module direction to connect to space inside

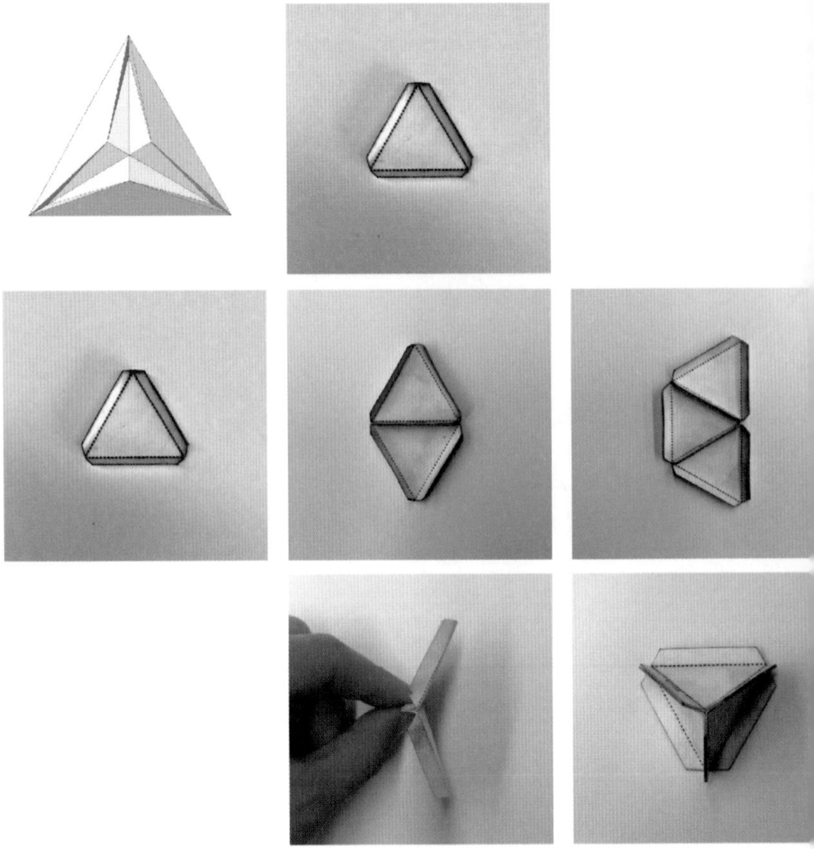

Unit Assembly

Physical Model Module Assembly

Render Assembly

Render of final installation

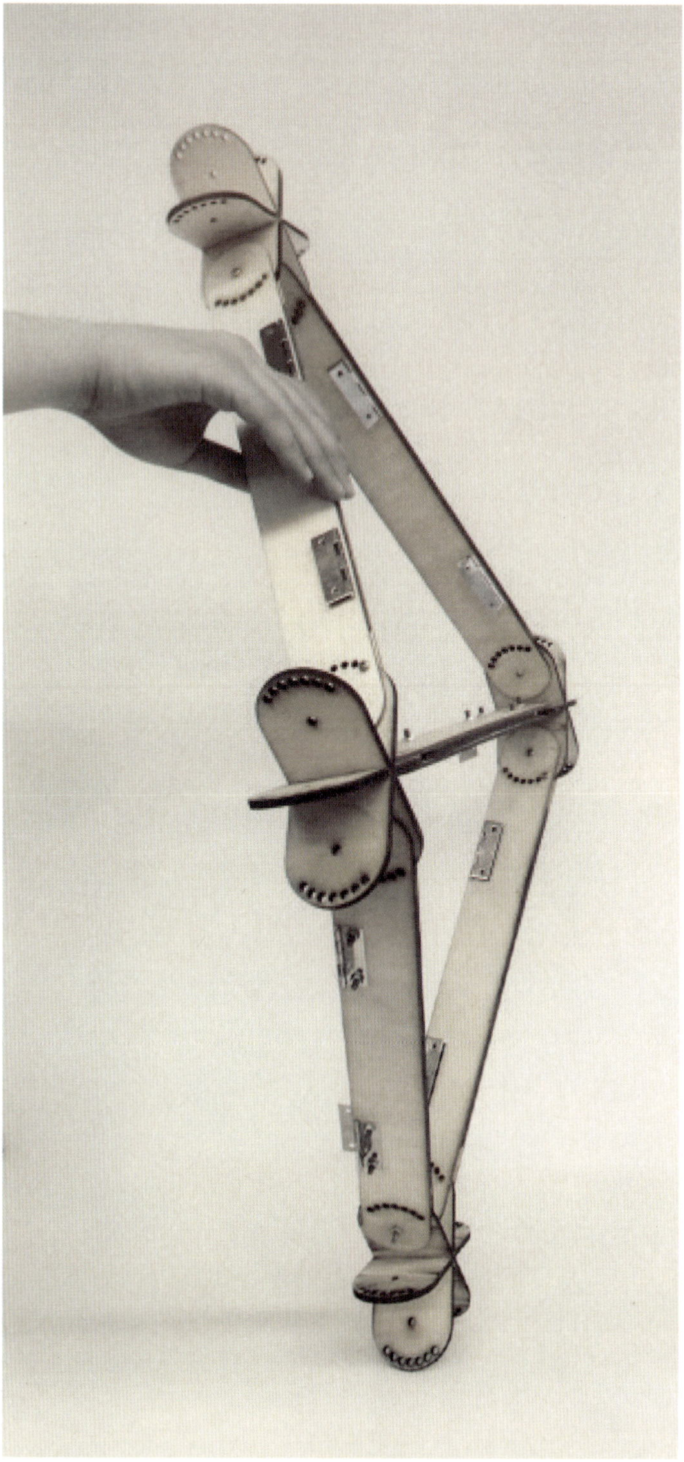

Models

Physical Model

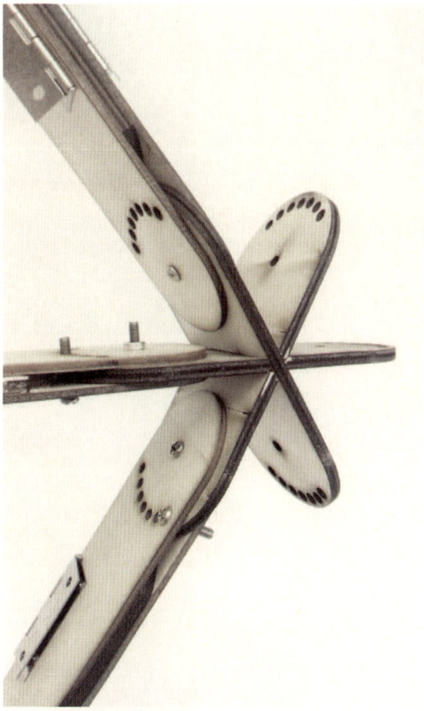
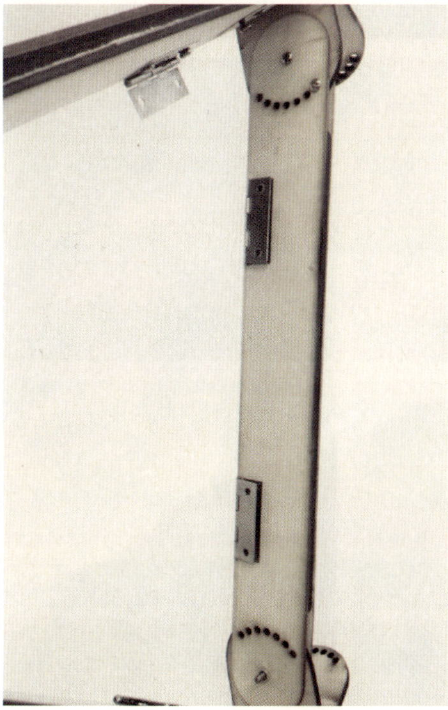
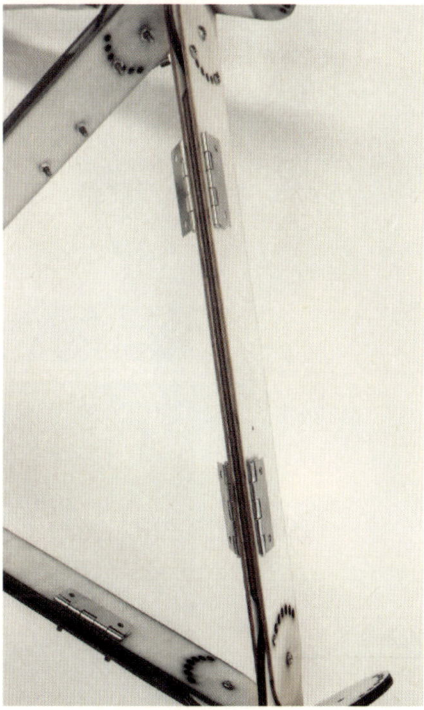

Models

Physical Model

Models

Physical Model

Responsive Kinetic Structures Based on the Tropic and Kinetic Movement of Plants

Problem-Driven Design

In this project, the biomimetic approach followed a problem driven method, in which an issue to solve is developed before sourcing a solution. This issue was focused on how to develop a design that could collect water during heavy rains for rooftop agriculture. The solution needed to be passively stimulated and have different stages of form which could react to weather conditions. We decided that the materiality and form would be intrinsic to allow for formal change.

Turgor pressure which builds up in plant cells allows for the angle of the stem to bend towards the light throughout the day. These cells change in their shape due to varied pressure, which allows the sides to contract and expand, affecting the angle. Following this logic, the research explored patterns that would allow for changes due to pressure. The selected pattern showed the strongest ability to retract back to an original state due to its form.

Materiality was tested along with the pattern, to ensure the movement transferred as the model scaled up. The final model was formed through a relationship of contracting mechanisms next to rigid modules that had flexible seams to allow for movement.

The application was intended to be useful for extremes in weather, mainly rain during the summer months for the sub-tropical climate in Shanghai. The mechanism is triggered when a heavy load is applied from above, which feeds into a cistern for water storage. The prototype lends itself towards other installations where a passive mechanism can be triggered due to extremes in loading such as wind, water, or snow.

Models

Final physical model

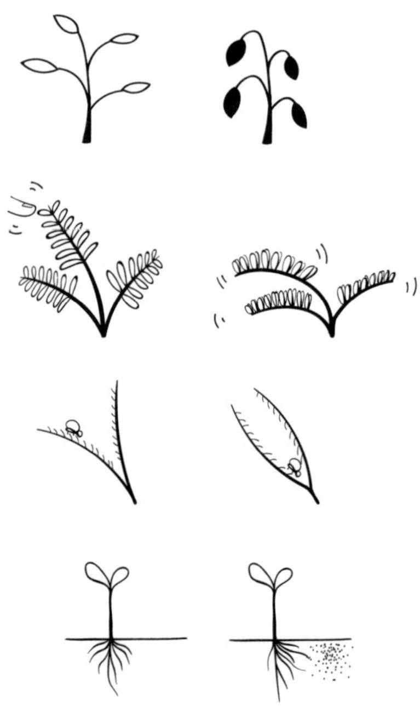

Responsive Movement Patterns

Current Page:
Plant movements according to age, touch, seed dispersal, and water source seeking mechanisms

Opposite Page:
Paper model studies developing patterns that allow for movement

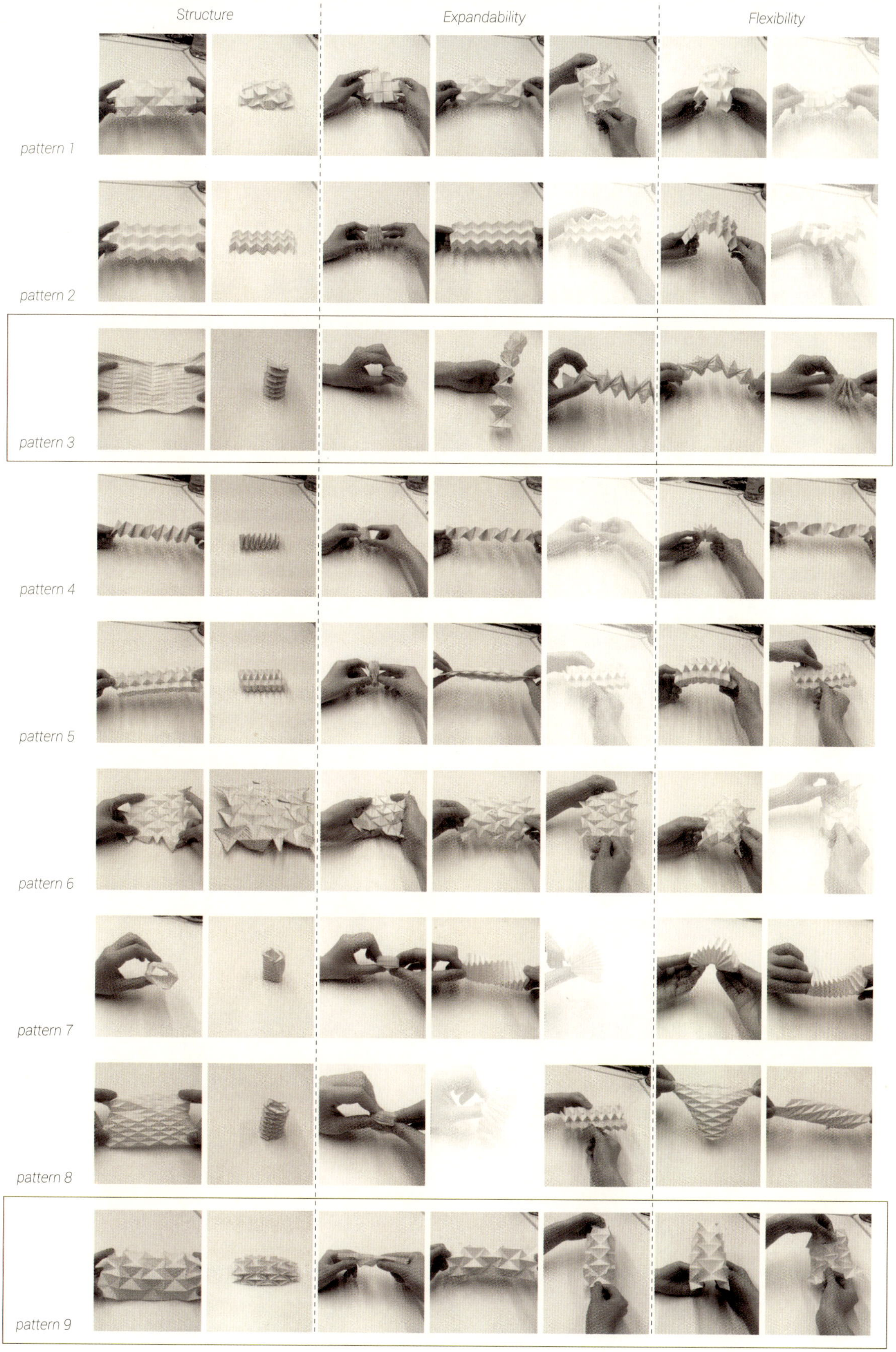
Structure
Expandability
Flexibility
pattern 1
pattern 2
pattern 3
pattern 4
pattern 5
pattern 6
pattern 7
pattern 8
pattern 9

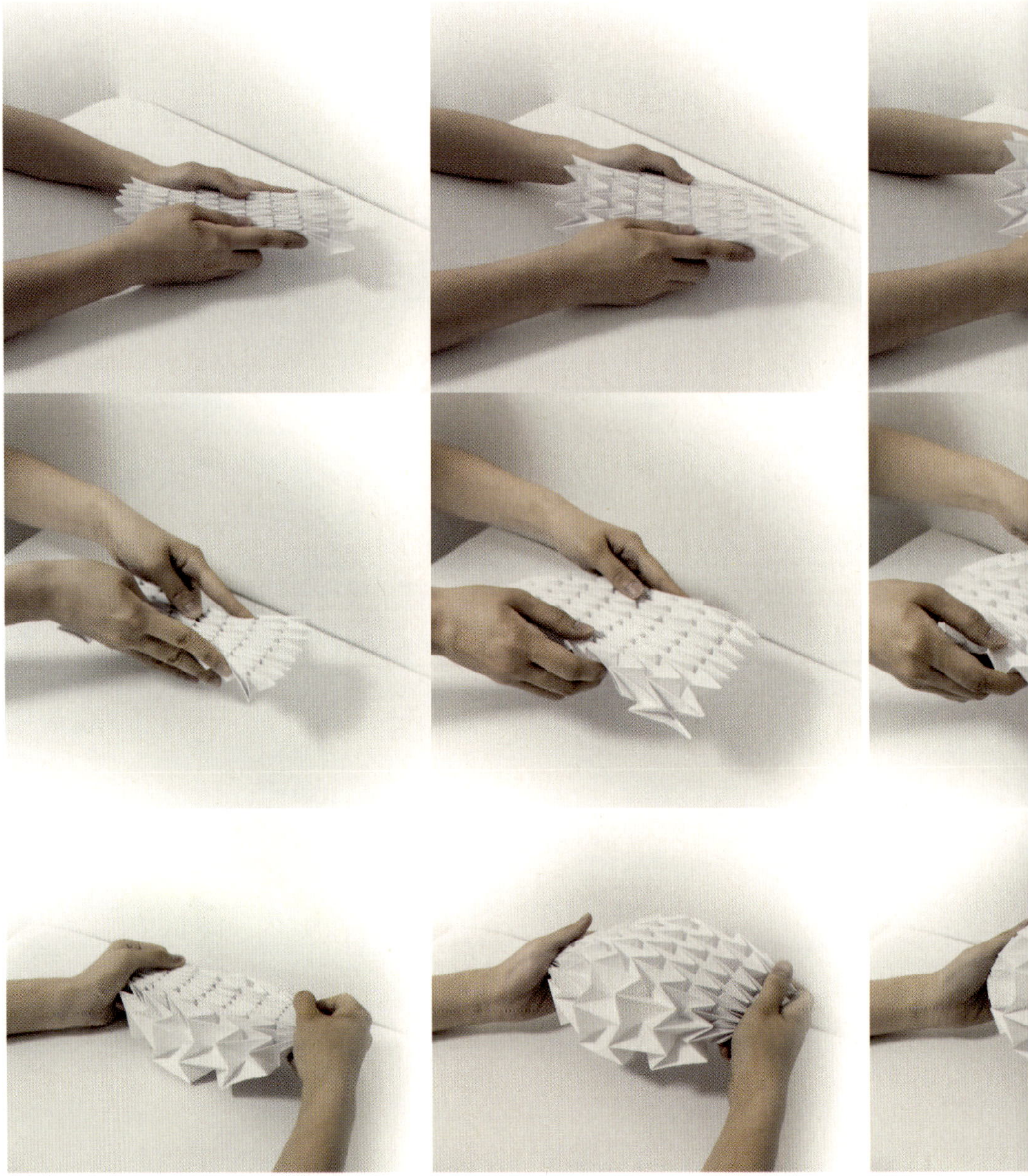

Movement

Selected pattern range of movement

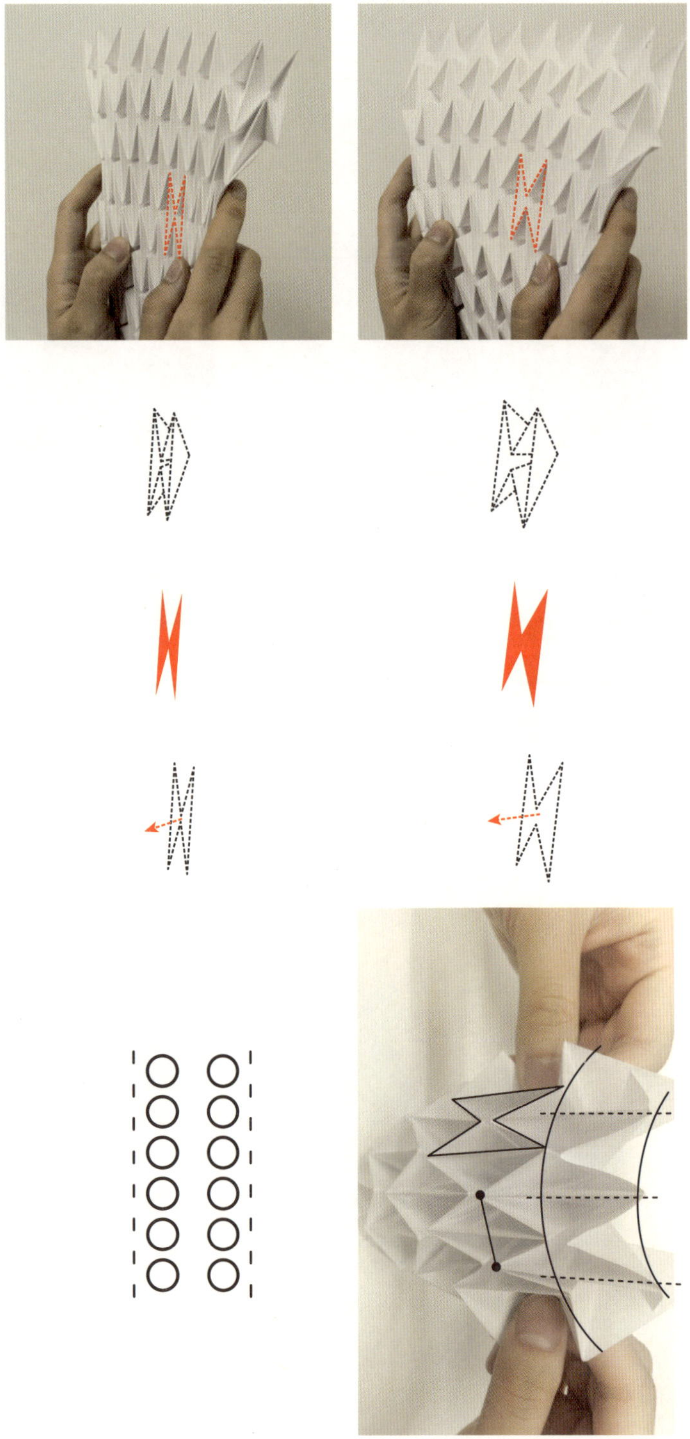

Patterns Isolated

Reduced paper model studies to explore a single modules movement

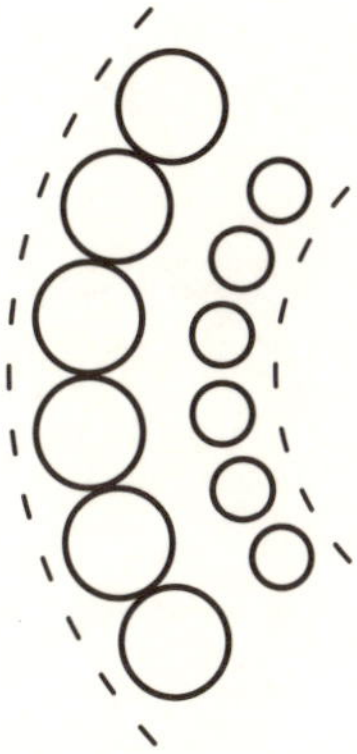

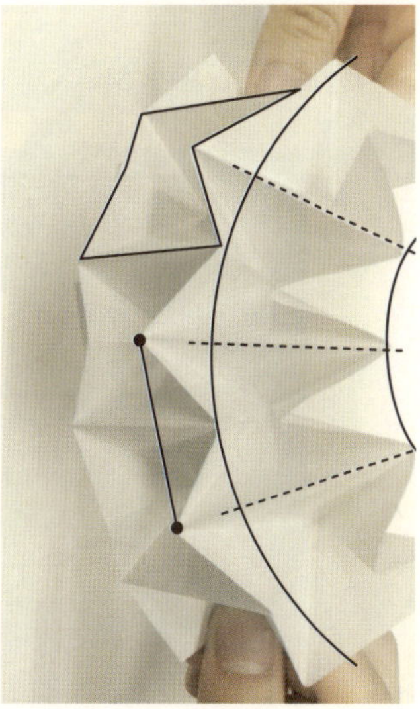

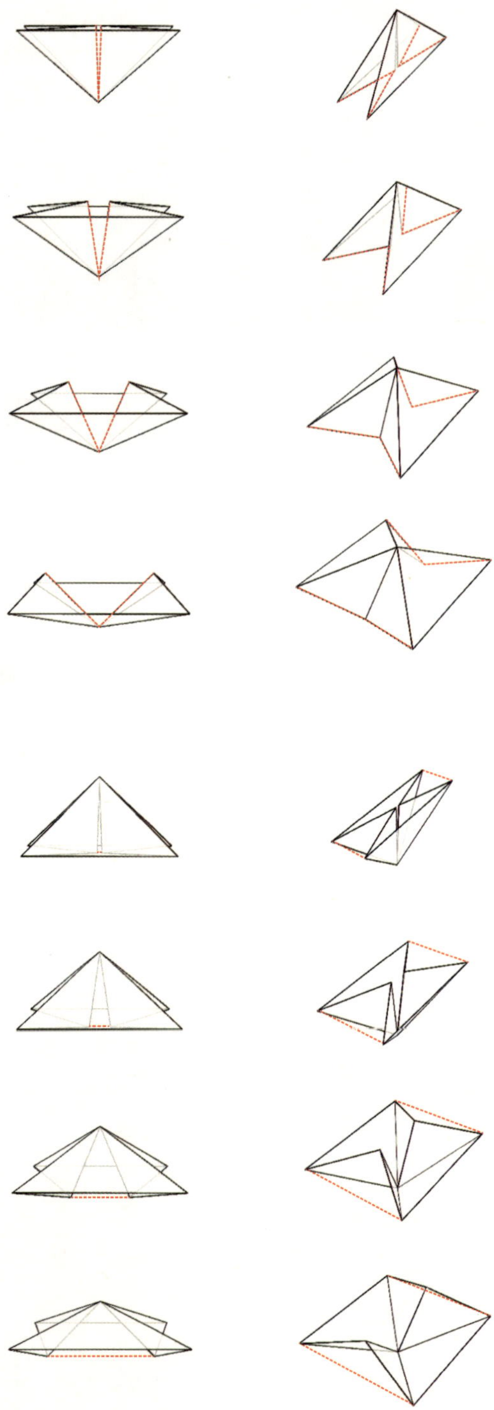

Module Movement

Reduced paper model studies to explore a single modules movement

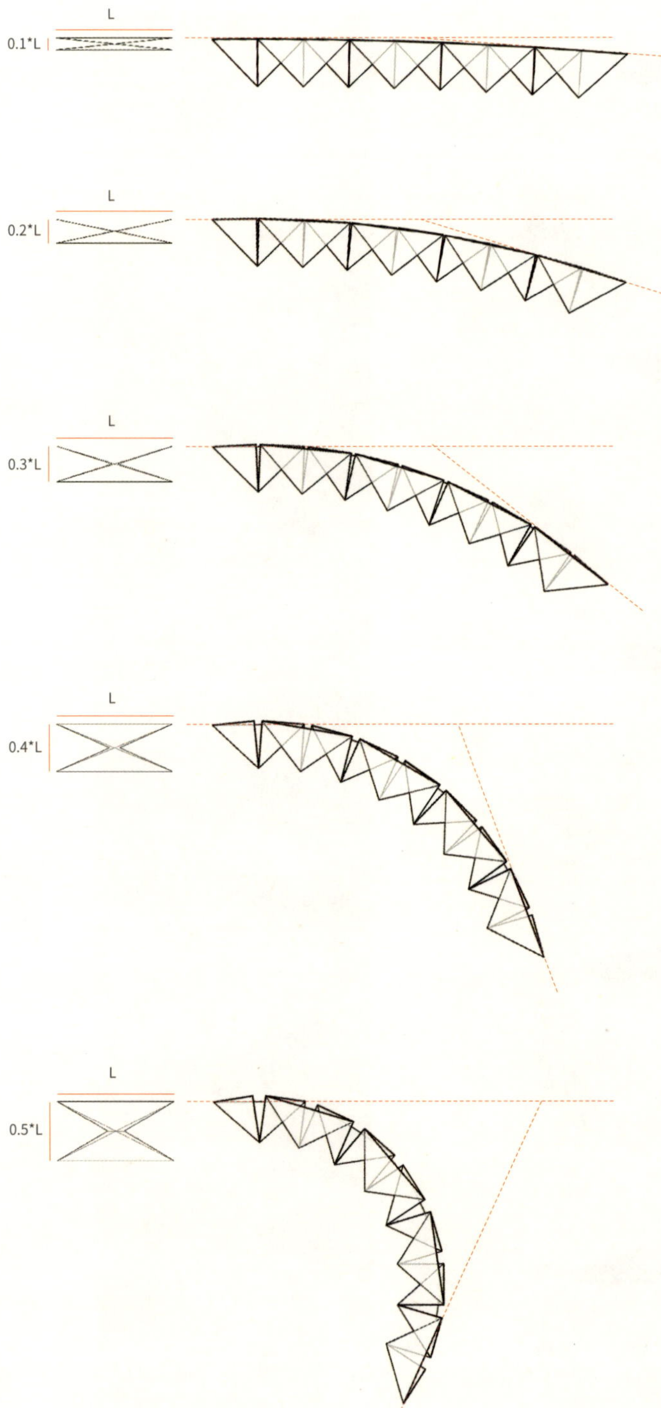
L
0.1*L
L
0.2*L
L
0.3*L
L
0.4*L
L
0.5*L

Actuator Evaluation

Conceptual exploration of dynamic force of rain or wind on folded pattern

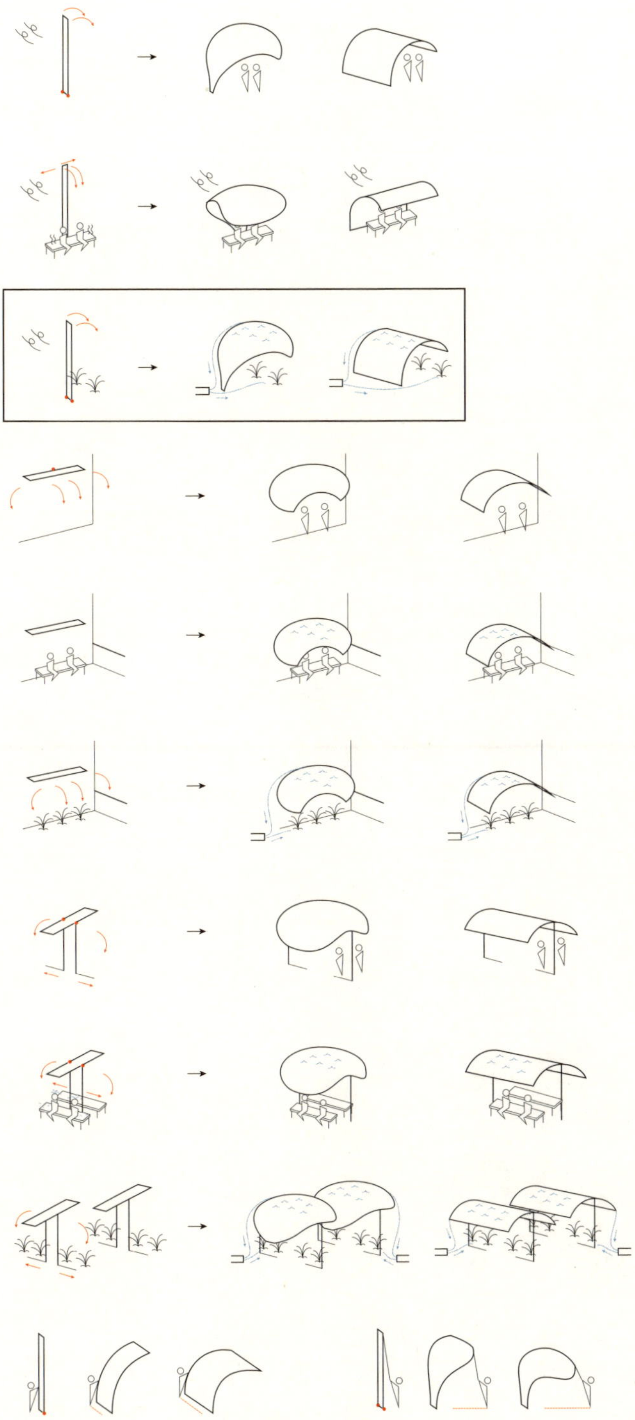

Scenarios

Prototype concepts showing conditions in which system could be deployed

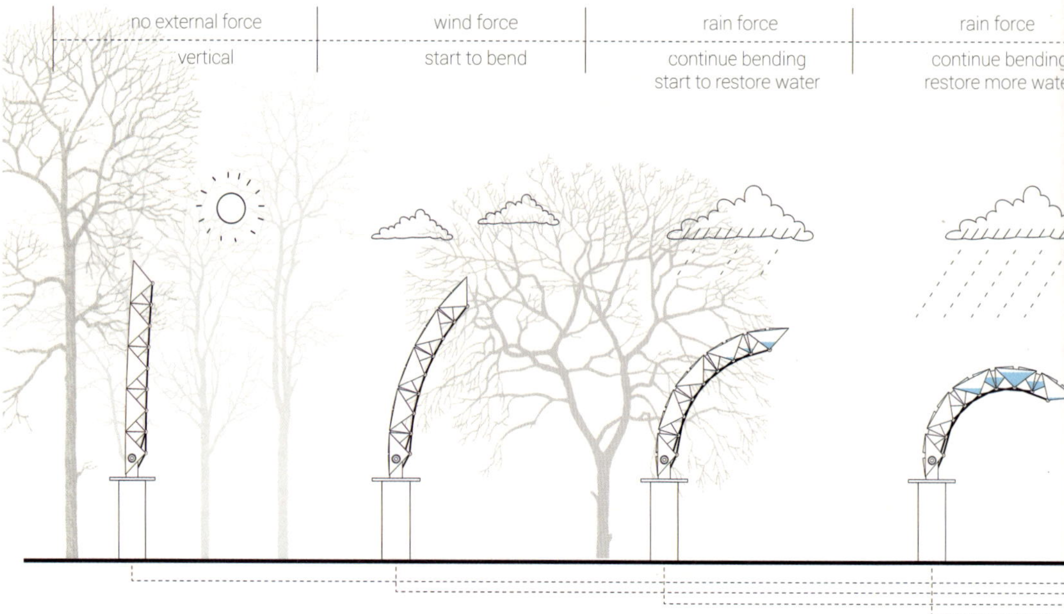

System

Environmental Responses: Rain Collecting And Irrigation System

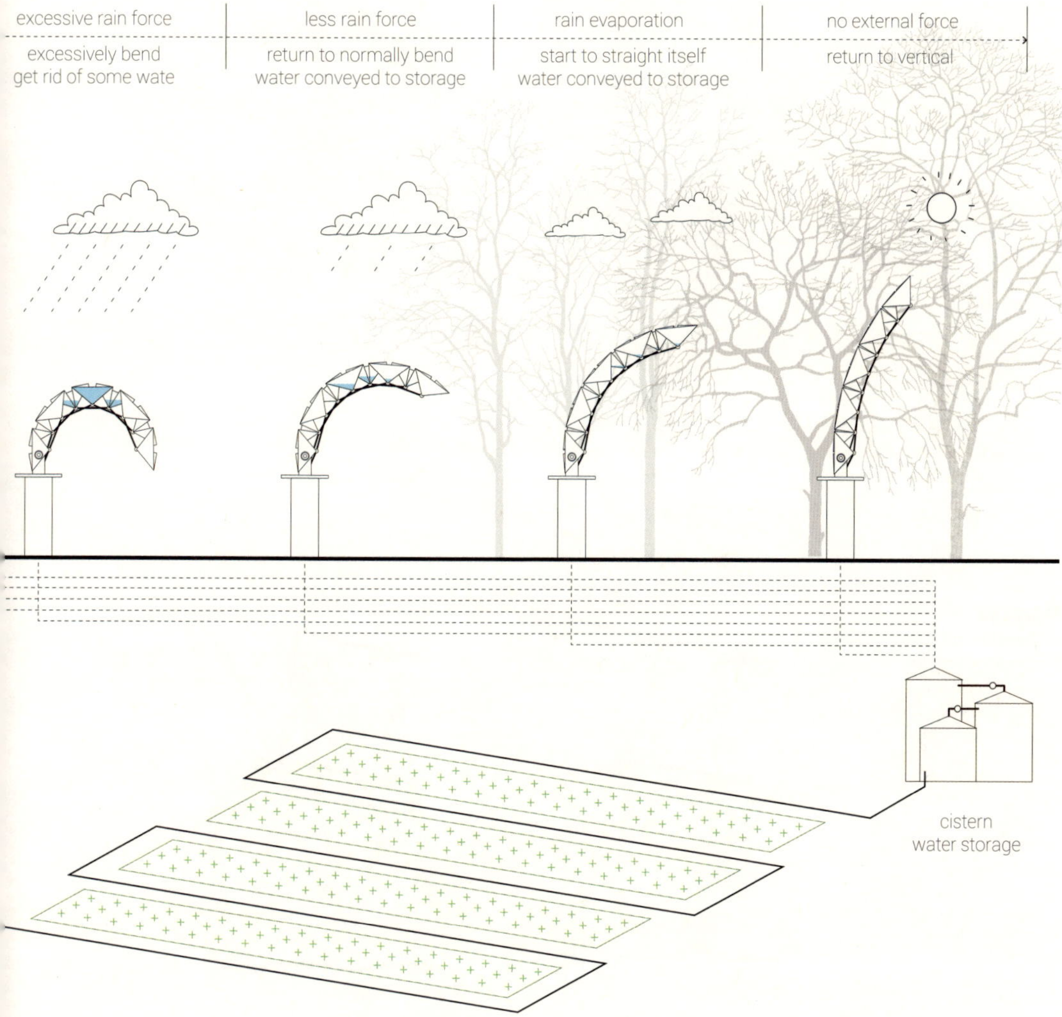
excessive rain force
excessively bend
get rid of some wate
less rain force
return to normally bend
water conveyed to storage
rain evaporation
start to straight itself
water conveyed to storage
no external force
return to vertical
cistern
water storage

Structure Composition

Left Image
The surface is used to enclosure a space for water collection, so the material needs to be water-proof and durable. It contains four frames to retain the whole shape. The surface can be disassembled for cleaning.

Right Image
The frame is the skeleton of this prototype. Each frame is connected by a rotating joint component.

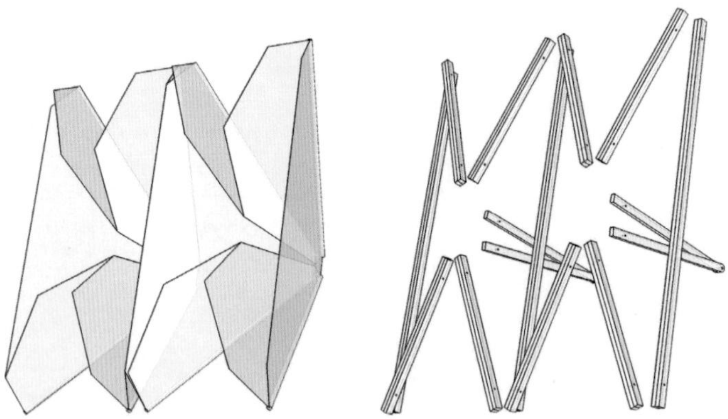

Structure Composition

Left Image
This prototype motion requires an energy to allow the structure become vertical after it bends. The spring can offer this energy as it restores elastic transformation energy when lengthened and released this energy when external force disappears.

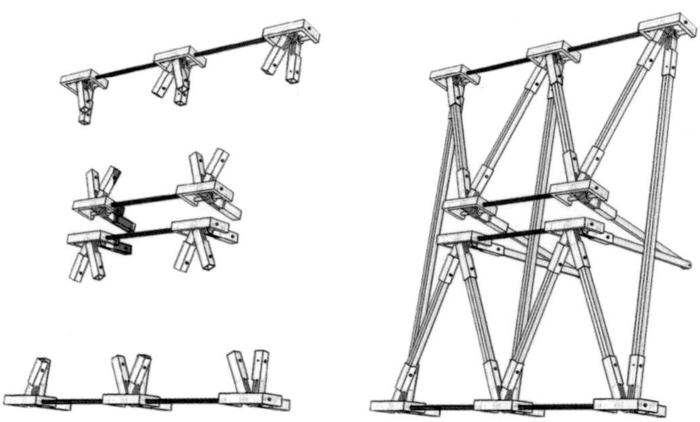

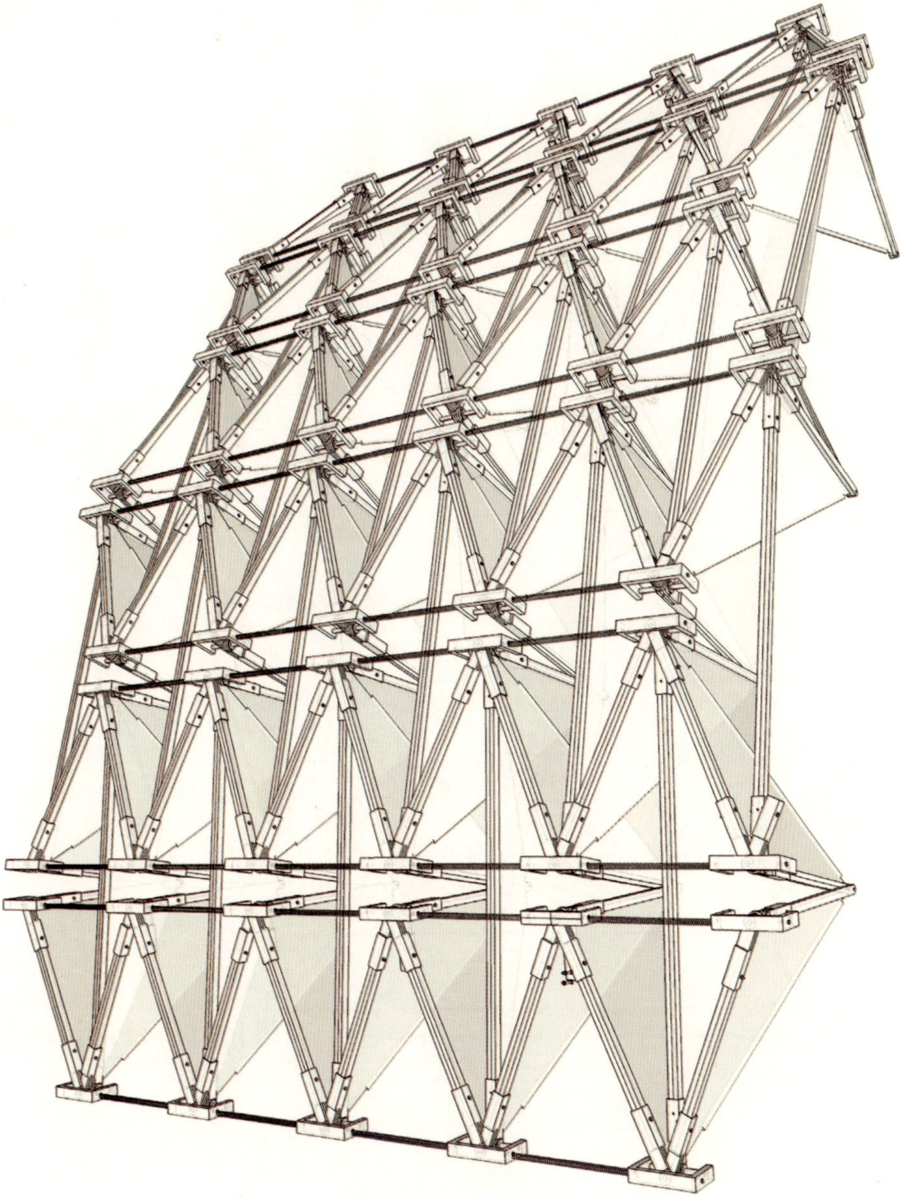

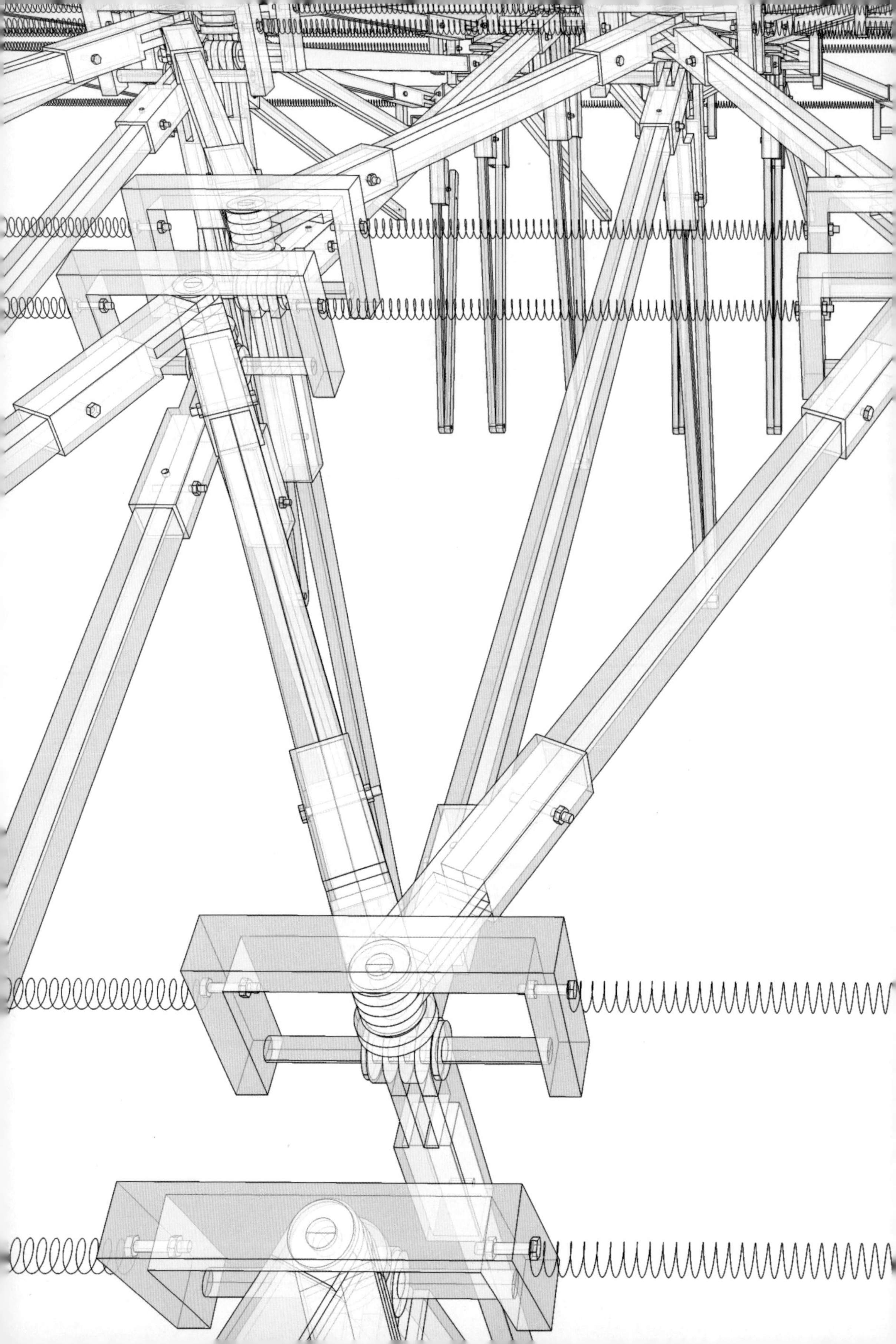

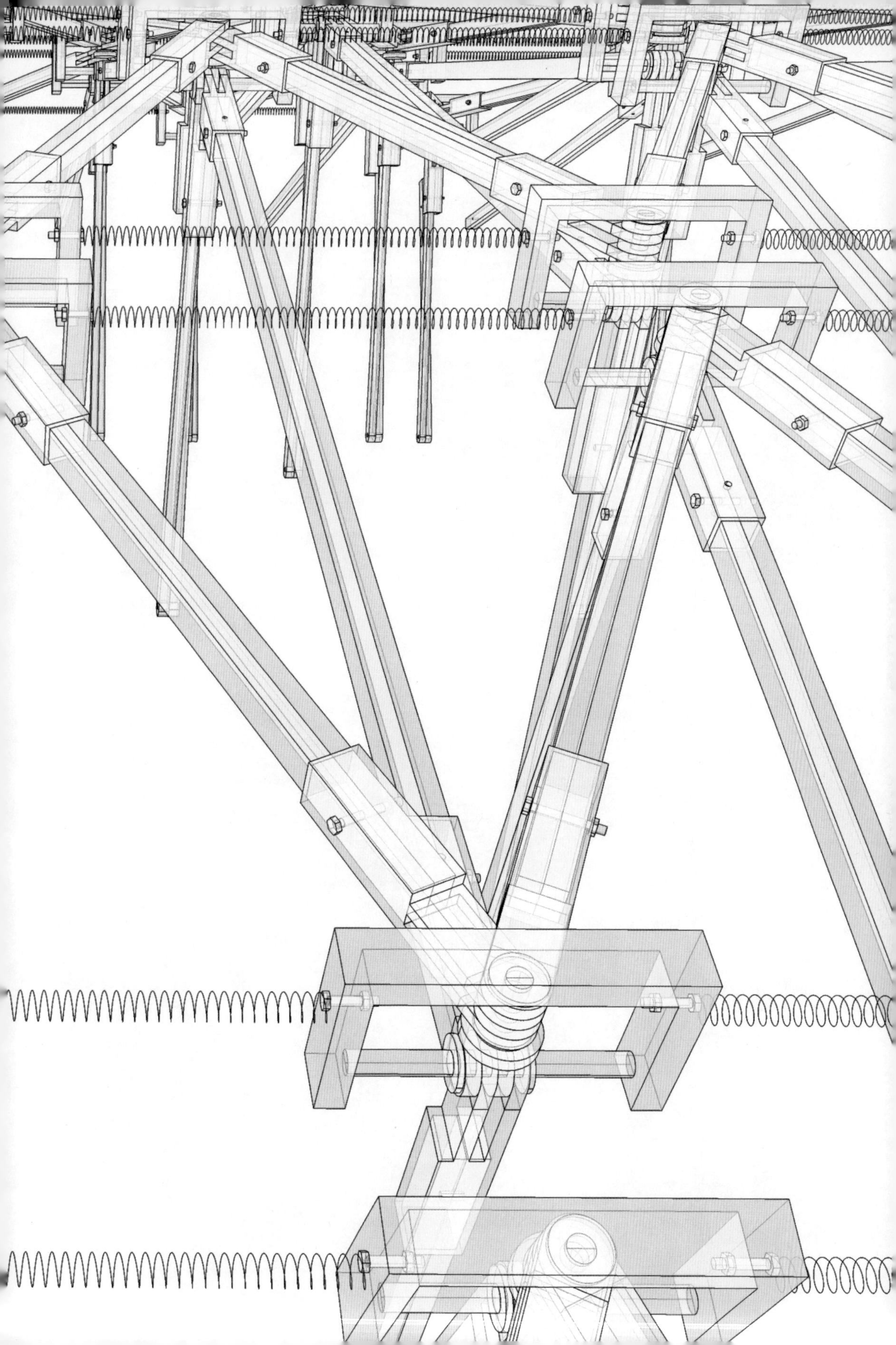

Design Education Methods

As part of the teaching agenda at BiDL, students are required to engage in projects that cover multiple scales of design with nature. These projects range from the urban scale, where ecosystems and environmental phenomena are analyzed, down to the microscale, where the origins of material properties are studied.

The principle that ties these studies together is a holistic vision that helps students understand how to evolve from an unexpected discovery in nature into a matured design proposal. This multi-scalar approach also allows for an understanding of the self-similarity of natural systems through observations.

To provide some insight into these principles, three projects have been curated to express the complexity and study typically explored at the BiDL: Nature NiDi at the city scale, the Guard House Seminar at the architectural and human scale, and the Biomimetics Seminar, at the micro scale. These projects cover aspects of design through infield exploration, technical evaluation and social and environmental responsibility.

Fern Shape

Blechnum spicant ferns are considered to have natural shape-memory properties as they can return to different states depending on their state of water intake.

Urban Scale Design

Nature NiDi is a conceptual research project which sought to integrate Shanghai's park systems with a public platform. The outcome of this study was to develop a means that allowed citizens to reconnects with urban locations where nature is surviving and thriving in its non-manicured, untamed, undomesticated character. NiDi was envisioned to become a platform that mapped, documented and evaluated sites of urban ecology, as well as footprints of biodiversity. Currently, the project is still under development.

The genesis of the idea stemmed from a connection between biology and the urban scale engaging the biophilic needs of people in the city. The term we use "biophilia" was first used by Erich Fromm to describe "connections that human beings subconsciously seek with the rest of life." Biophilia suggests that, in spite of an ever-increasingly urbanized world, humankind is inseparable from the natural environment, and that 5,000 year-entities like cities can't replace millions of years of intrinsic connection between man and nature.

Presented in the following chapter is the background research for this project which includes a selection of documented parks. As one of the first steps in the work for this project, students were brought into the city's parks and taught how to evaluate natural urban areas. During these site visits, experts and teachers joined in observing the species and ecological systems, which helped to determine key aspects unique in each park.

The following images and maps depict everyday life, key features, materiality and visitors in each of these contexts. This narrative and imagery were used to set up the catalog of information for the project.

Urban Green

Huangpu River, Shanghai

Mapping of key urban green areas of Shanghai for Nature Nidi platform

Houtan Park

Houtan Park is one of the more famous parks in Shanghai, known for its beauty, riverfront adjacency and revitalization of a former brown field industrial site. The park was developed from a concept of restoring the Huangpu waterfront and to remediate the polluted river water through phytoremediation. The park has areas that promote visitors to experience nature through a series of wetlands, ecological flood control zones, reclaimed industrial structures and urban agriculture.[1]

The site for the park is along the waterfront edge of the Huangpu River. The length covers over 14-hectares and connects to the Shanghai Expo site. The site was previously a shipyard and steel factory supporting the demands of the river. However, in more recent years, was converted to a landfill yard for industrial materials and scrap.[2]

In documenting this site, we took note of the parks biomes used for water cleansing process and indicated how they were connected to the experience of the visitors. In many areas, direct access to edges and transitional zones between water bodies and forests were linked with pedestrian paths and entry points. Through this composition, the engagement with the natural space welcomed urbanites to see relationships that occur in nature and restoration practices necessary to bring life back to the area.

Houtan Dock

Waterfront dock at Houtan Park, Shanghai China

1 "Houtan Park" Landscape Voice. September 22, 2013.
2 "WATER PURIFICATION_ Shanghai Houtan Park: Landscape as a Living System." Encuentura. November 24, 2014.

Houtan Dock

Views of the Houtan Park landscape, Shanghi China, Fall 2014

Luxun Park

Luxun Park is an authentic Shanghai park which serves adjacent residential neighborhoods in an area near Hongkou Stadium. Historically, the park was an old British shooting range which was converted into a local park and dedicated to the Chinese writer Lu Xun as he lived in the nearby area.[1]

The park provides key features of allées for strolling, a boating area, mahjong, dancing and tai chi areas. The parks connection to nature is curated through a typical urban landscape, however, there are areas of lush overgrown shade and understory foliage that has developed since 1896. In terms of connecting visitors with the landscape, this park is successful as it provides an area of reprieve and respite in the city. The park has been selected as part of the NiDi mapping list, as it provides connection to the concept of biophilia and need for nature within the city.

Although there are limited areas to observe ecological processes, the age of the park has allowed the green space to act as a landmark for visitors. Many birds, insects and fish have inhabited the park. This suggests that even in parks where the landscape is cultivated, natural relationships can survive to benefit the adjacent urban context.

Luxun Park

Man stretching on tree in Luxun Park Shanghai China

1 "Luxun Park." Beijing Visitor - China Travel Guide. December 13, 2017.

Luxun Park

The park features many landscape amenities such as calligraphy space, river seating next to the large lotus and meandering paths to enjoy the water.

Binjiang Forest Park

Located in Gaoqiao town, Shanghai Binjiang Forest Park is the largest suburban forest park close to Shanghai. The park presents a combination of features that promote natural habitats, protected areas and recreational zones. The concept of the park focused on displaying wild and natural habitats.

More than 200,000 species of trees and bushes are planted within the landscape that contribute to robust zones of forest, field, meadow and waters edge. Visitors engage in these features through wooden paths, playgrounds, trolley and camping areas. As the park is adjacent to the Huangpu River, the Yangtze River and the East China Sea, there are views of nature, the city and the sea.

The park is a successful example of a significant ecological infrastructure for urban areas. It holds habitats for migrating birds that use the areas as a flyway connecting to Chongming Island. This park is a significant urban attractor for visitors to engage with nature from the adjacent residential areas, as well as Shanghai.

Binjiang Water Network

View of one of the many waterways at the park, which contributes to the ecosystems of the landscape.

Binjiang Water Network

Views of the landscape features in the park. From open fields for camping to the connected waterways, the park is large and extensive for people and varied ecologies.

Sanhao Park

Sanhao Park was built by the teachers and students in the campus near Siping Road as part of Tongji University. There are trees throughout the park, as well as the primary distinguishing feature here, the south bamboo forest, in which many teachers and students read and exercise in the early morning. The lake is in a scenic area and is used for scientific research projects focusing on water quality management.

The park provides a strong connection to nature and natural processes through water and forested areas that are free to develop within their context. The visitor paths are minimal and bring viewers close to the lake edge and deep within the forest areas.

This park was selected as part of the research as it provides an educational model for natural processes. Its frequent connection to class work for scientific study, as well as leisure and holiday activities on the campus, make it a significant urban green space.

Sanhao Park

Tree extended over the waterway in Sanhao Park, Shanghai China

Sanhao Features

Dock bridge over the main waterway.

Gongqing Forest Park

Gongqing Forest Park is located in the northeast of Shanghai and is adjacent to the Huangpu River and Jungong Road. The park covers a total area of 131 hectares. The park is comprised of two main zones, the north and south gardens.

The north garden is known for its botanical scenery, with nearly 200,000 trees and over 200 species of plants. The main feature areas of the park are the hills, lawns, lakes, rivers, forests and bamboo groves. The South Garden is a theme park focused on bamboo culture with amusement areas for adults and kids.

In the interest of urban attractors, the park as a whole covers a large space and provides a strong green landscape for natural and commercial use. The lush northern portion of the park allows for nature to establish itself organically through integrating wetlands, habitats, and forested areas. The use of the park invites visitors to step off the curated path and enjoy the landscape freely. In terms of the biomes and ecological aspects, the park would be beneficial in connecting people with a range of species of plants and birds throughout the year. The arrangement and procession for the visitors set up a close engagement with nature, ideal for infield expert tours.

Gongqing Walk

Visitors enjoying the spring and a walk down the park.

Gongqing Forest Park

Varied landscapes in the forest park

Human Scale Design

In the Guard House Seminar, techniques were introduced which explored how to use environmental conditions for designing habitats. There are many resources and tools that facilitate the study of climate and often times students must develop technical information to prepare a design. As an example, when considering aspects of humidity and temperature control, intensities of either factor can vary the outcome of a habitable or inhabitable space.

The project aimed to address issues of environmental comfort found throughout the guard houses at Tongji University. A guard houses provides shelter for the staff which monitor the entries to the campus throughout the day. Typically, these enclosures provide basic shelter from wind and rain but don't include thermal comfort or amenities such as water and restrooms. The main issue associated with these spaces was how to control the climate on the interior with minimal intervention and reliance on systems that would require additional energetic support. This problem is relevant throughout all of Shanghai, as many facilities, institutions and corporations have guard houses with the same limited amenities.

The site selected was the guard house located at the main gate in the southeast area of the campus. It is adjacent to Siping Road, the main road of access to the university and where human traffic is consistently high on a daily basis.

The students developed their projects following three phases. The first phase focused on data collection and framing the issues to guide the design. Students interviewed and frequently visited the house to establish relationships with the staff and understand the needs from an in depth perspective. The second phase required students to research biological systems and source a potential solution. Students worked through a series of concepts to reach a final design that worked for the site, scale and needs of the staff. In the last phase, students developed full applications for the guard house and were asked to articulate them as an architectural intervention.

Guard House

A Tongji University guard house on the north side of the main entrance gate at Siping Campus

Research led by Pius Leuba.

Comfort Zone

Image Top:
Psychrometric Chart
Shanghai China
Jan. 1st - Dec. 31st
Time: 00:00-24:00
Ecotect: Weather Tool

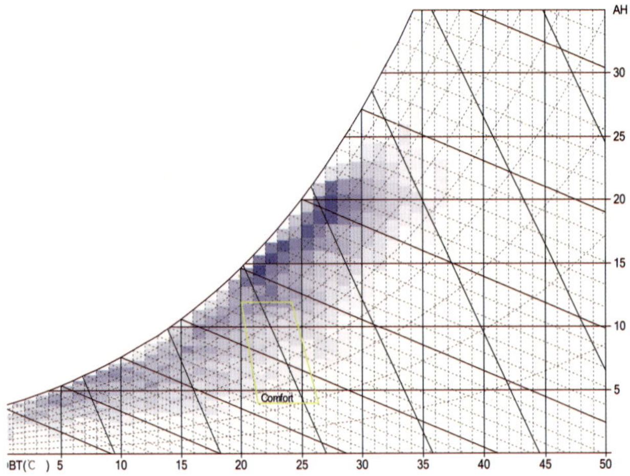

Annual Conditions

Annual weather trends produced from Ecotect studies

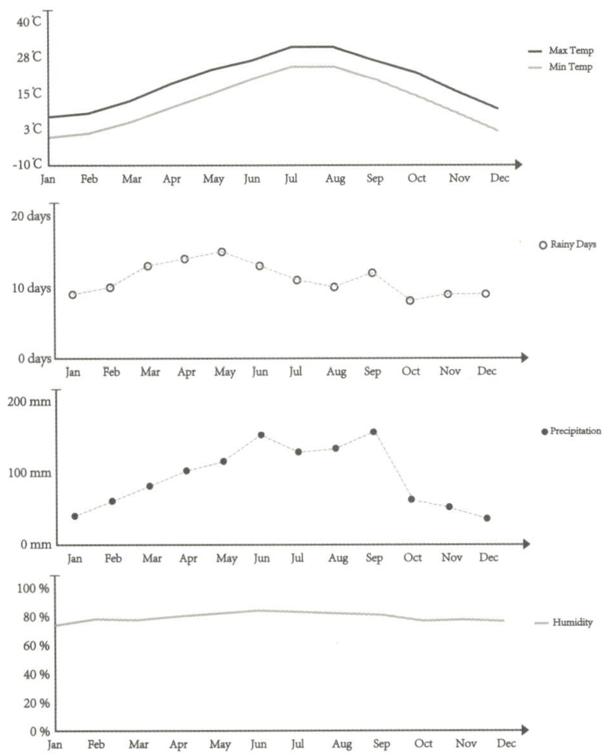

Climate

Due to the humid subtropical climate (Köppen Cfa) in Shanghai, the temperature changes drastically throughout the year. During the summer months, the temperature can swell to over 30°C in July and August and in the winter months can reach about 3°C during December and January. In terms of precipitation, total amounts of rain equaling 100mm can be reached from March to September. During the peak period of the monsoon in June and September, total rainfall is about 150mm. This high precipitation leads to high humidity which consistently puts Shanghai in high humidity zone throughout the year. On average the relative humidity (RH) is about 80% per month while in the peak months of June and July, the RH is over 80%. This percentage changes the level of the temperatures to fall outside of human comfort zones.

Challenge

After analyzing the weather and finding the environment too cold in winter and too hot in summer, the challenge was to study the conditions of the site to alter these effects. To begin, we studied the exposure of sunlight as the main source of energy for a passive means of altering the comfort level. Throughout the year we found that the guardhouse was only exposed to sunlight in the morning and afternoon, as sunlight was blocked at other times by surrounding buildings and landscape. For wind, we found that at the back of the guardhouse, the neighboring building blocked predominant winter winds which would increase cold weather. During the summer, the prevailing winds shift in direction and generated cross ventilation for the enclosure. From these studies, it was decided as one of the main issues each project should focus was how to find the best orientation for sunlight and summer wind to reach a specific comfort level.

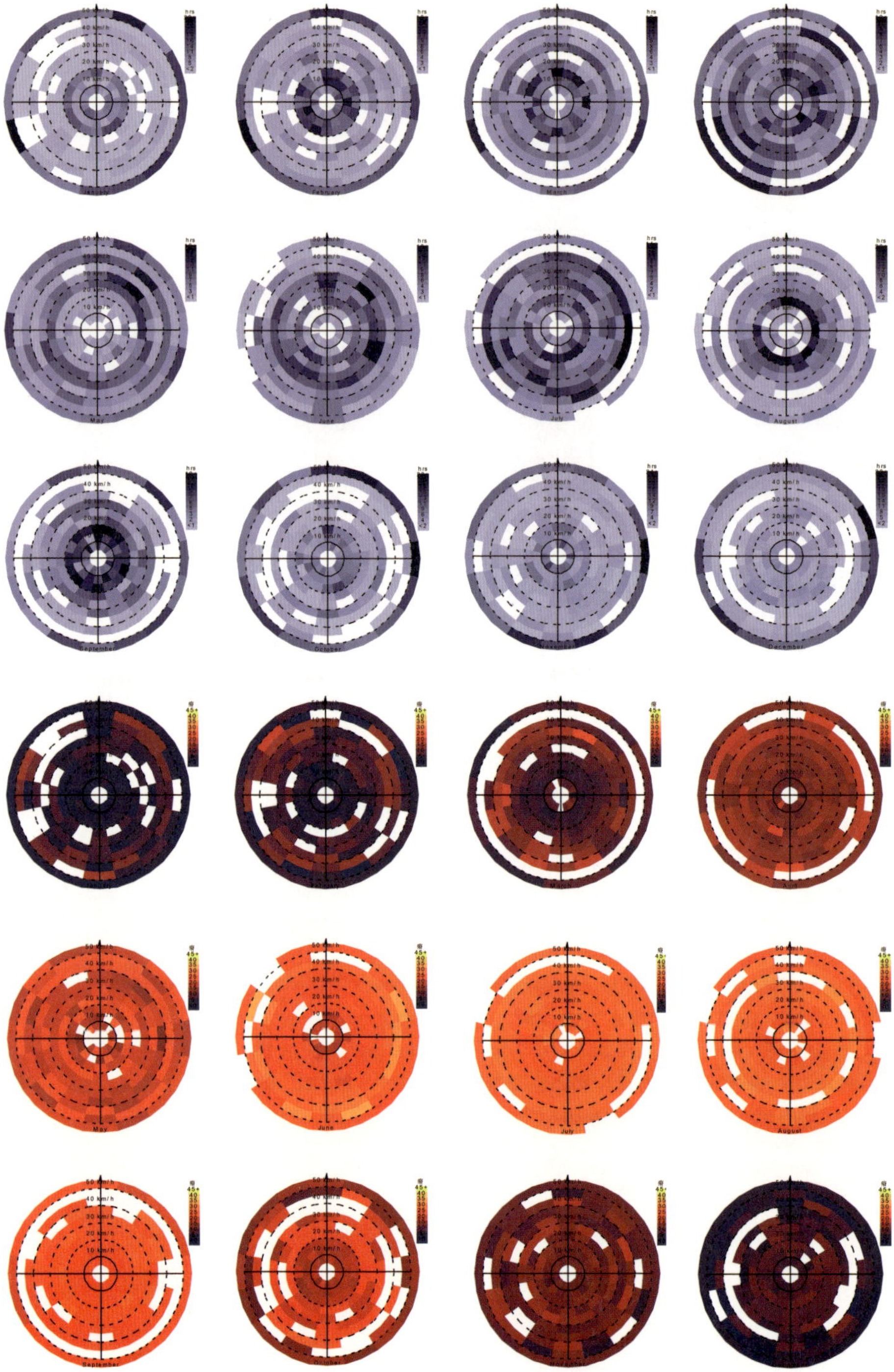

Spatial Challenges

To understand the patterns of use of these houses, students went on site and documented the workers throughout the day. From this shadowing exercise, there appeared to be three main conclusions: Problems of minimum space; problems of limited energy supply, and issues associated with limited resources.

Analytical Outcomes

In addition to the climatic research, we carried out an analysis of the periods of extreme weather. On average, there were approximately 46 days of weather annually considered hot in terms of human comfort. In opposition to this, there were 20 days of cold weather. In terms of precipitation, there were annually 5 days of heavy rain, 7 days of snow, and 7 days of typhoon weather. It should be noted that these ranges were supported with data sets from a series of years collected by weather stations, however through interviews, many said that they felt more uncomfortable in the summer and winter than the amount number of days suggested.

In addition to these findings and through observations, users of these shelters suggested that summer heat was a greater issue of concern, than the cold days. The conclusion from the analysis and interviews was that the design needed to address conditions of cooling for the high humidity and heat for summer days and improved layout.

Student Prototypes

The student work for this course can be found on the following pages. For each project, the biological inspiration used to determine a natural strategy for design is shown. Following this inspiration are concepts that address the environmental issues of the guardhouse. These concepts did not need to address all of the issues, but rather to consider an in depth strategy and how it could be designed and implemented with the existing architecture.

Shanghai Weather

Prevailing Wind Frequency
Monthly Charts from
Jan. 1st - Dec. 31st
Time: 00:00-24:00
Ecotect: Weather Tool

Prevailing Wind Temperatures
Jan. 1st - Dec. 31st
Time: 00:00-24:00
Ecotect Weather Tool

Self Shading

Tree Bark

"The Bark of trees keeps its surfaces cool by minimizing absorption of solar light and maximizing thermal emission."[1] Bark is a critical feature that helps to protect and mitigate heat gain and loss. Tree bark displays a paper-like structure and these layers are highly insulating, as they allow for trapped air to occupy spaces in-between. Trees with more of a rough surface, have been studied to show that these surfaces produce a lot of shadows which stimulate the convection of air, that moves heat away from the surface. In addition to this, the form of trunks, minimizes the surface with respect to the volume. For this reason, comparatively little solar heat can come in from the outside.

Inspired by these functions, we developed wall systems that through their layers and pockets, could direct air flow and promote heat absorption. The areas of application could be to the existing exterior walls to reduce the heat gain through mass during the summer months. In addition to this, convection was considered to occur through the cool side through internal cells, to allow for constant air flow on the interior spaces.

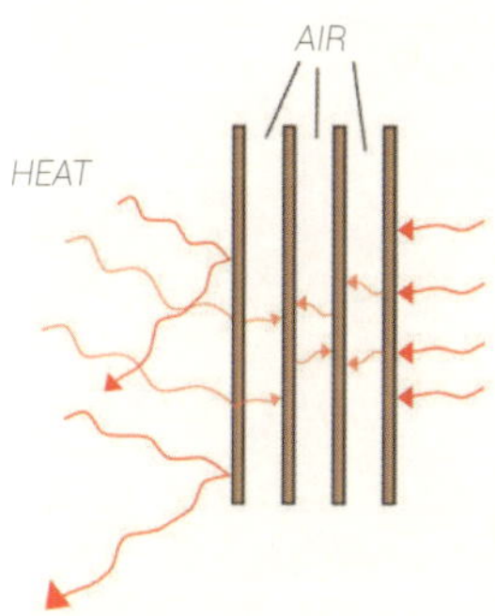

Bark Studies

Opposite Page:
Tree Bark Photo

Current Page
Top Image:
Concept Diagram of a bark inspired wall system

Axon views of wall system showing cells for air build up

Student Team
Sheng He , Kan Liu, Junying Li

1 Henrion, Wolfgang, and Helmut Tributsch. "Optical Solar Energy Adaptations and Radiative Temperature Control of Green Leaves and Tree Barks." *Solar Energy Materials and Solar Cells* 93, no. 1 (2009): 98-107. doi:10.1016/j.solmat.2008.08.009.

Light Filtration

Succulents

Succulents typically can be found in climates where there are extremes in temperatures which requires them to maximize energy absorption and store this energy for long periods of time. Shown here in the *Fenestraria aurantiaca* plant, there is a distinct light filtration mechanism developed for this purpose. The top morphology of the plants' skin is more transparent than the rest of the plant to support light filtration.

Sunlight enters the tops of the plant through a series of oxalic acid crystals. This light is diffused and reflected internally such that the sunlight reaches the grains of chlorophyll that are distributed around the sides and the bottom of the leaf.

The design inspiration from this feature explores a roof system that would use light as an energy resource and diffuse heat gain.

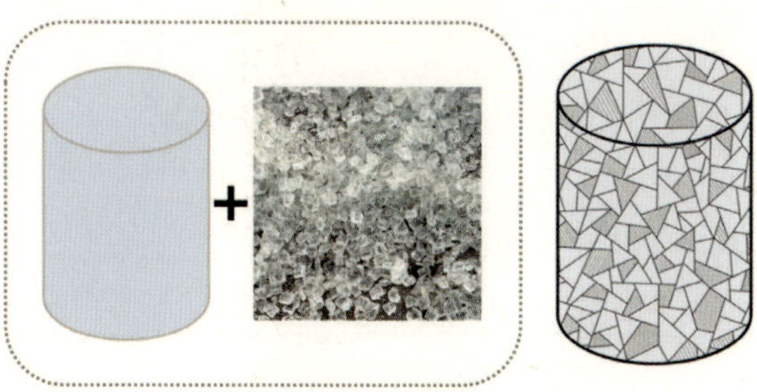

Succulent Studies

Opposite Page:
Image of *Fenestraria aurantiaca* or "Baby Toes" or "Window Plant"

Current Page
Top Image:
Concept diagram of sunlight absorption in walls

Lower Images:
Material concept diagram for *Fenestraria aurantiaca* inspired ceiling system

Light Pillars

Succulent skylight concept

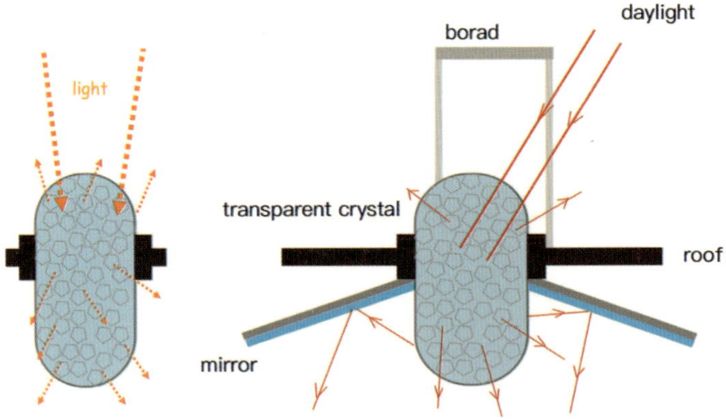

Roof Application

Guardhouse's elevation and section showing skylight concept

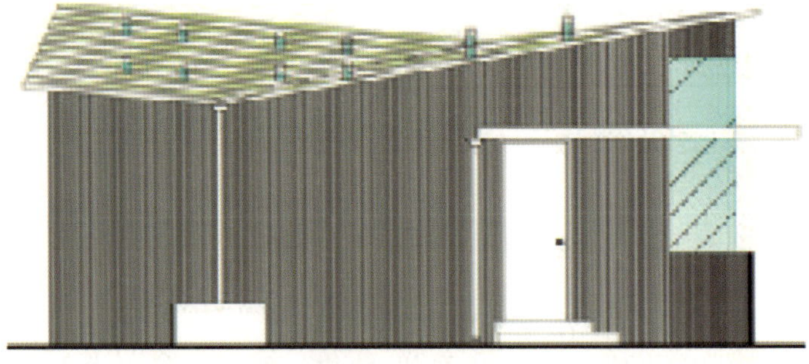

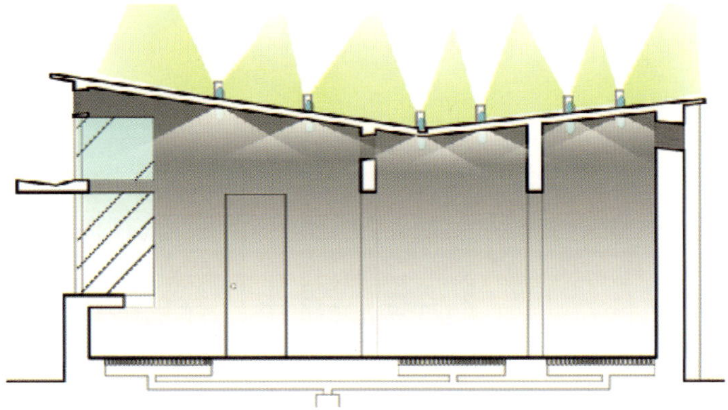

Light Concepts

Opposite Page: Images of light refraction and absorption during different times of the year.

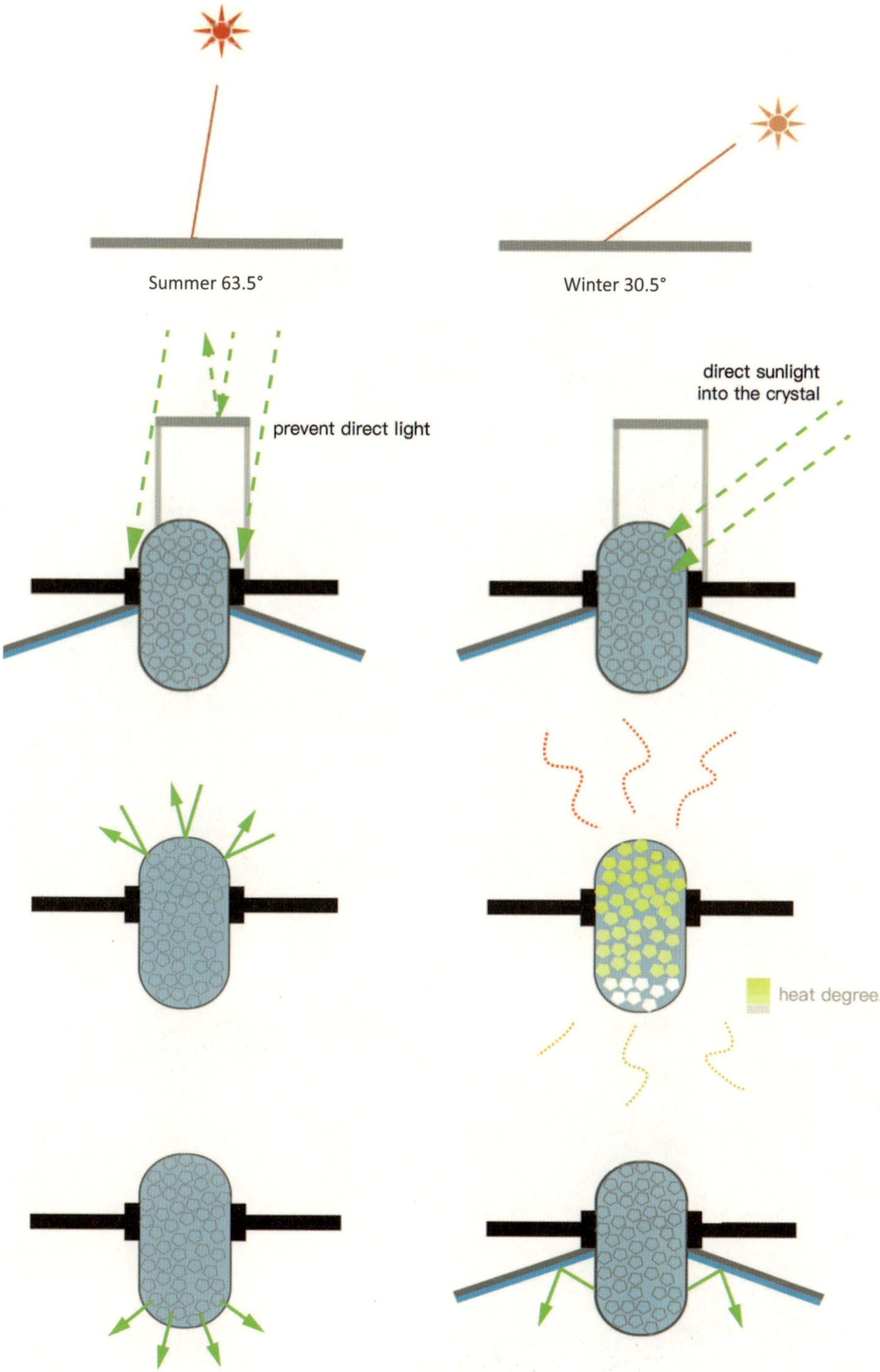
Summer 63.5°
Winter 30.5°
prevent direct light
direct sunlight
into the crystal
heat degree.

Hydrophobic Grooves

Bird Feathers

For birds, water is typically shed due to microscopic grooves on the surfaces of their feathers. These grooves generate a hydrophobic condition throughout their bodies which acts as a protection from wetness, but also in the transfer of cold from wet environments.

We developed a concept for buildings that could control the humidity which often attributes to the feeling of cold or hot. The concept generates a barrier between indoor and outdoor conditions by having air flow into the room from a lower intake source in the wall and exiting the room through an exhaust port at the ceiling. A filter with similar concepts of water collection as in feathers was located in the inlet to remove water in the air from moving into the room. Thereby reducing the effect of humidity on the interior temperature.

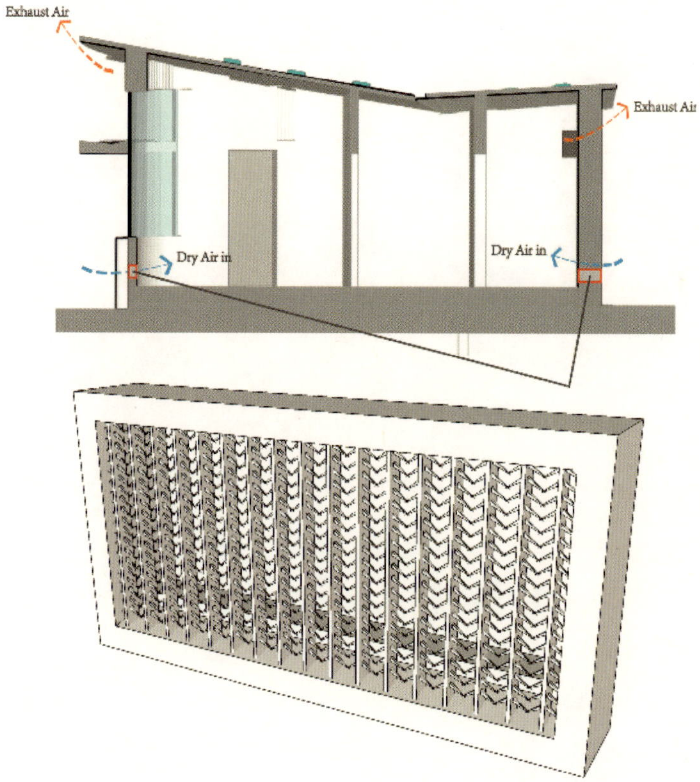

Feather Studies

Opposite Page:
Image of water bead on bird feather

Current Page
Top Image:
Section of guard house with air intake concept

Lower Images:
Concept prototypes for wall air system

Cooling Cells

Honeycomb

The honeycomb is a crucial part of beehives as their structures effectively regulate the temperature of the colony. Each comb is organized such that individually there is a delay in heat gain and direct sunlight. Thereby reducing the effect of temperature on the entire hive.

This concept inspired us to develop wall systems that could absorb intense heat from the sun without transferring it to the interior. The idea was to orient exterior walls in their environments directly facing the maximum sun angle during the summer months. The outer walls would be made up of "honeycomb" structure with mirrors to refract the light while still allowing for a view. The pattern would be porous to allow for prevailing winds to cross the space.

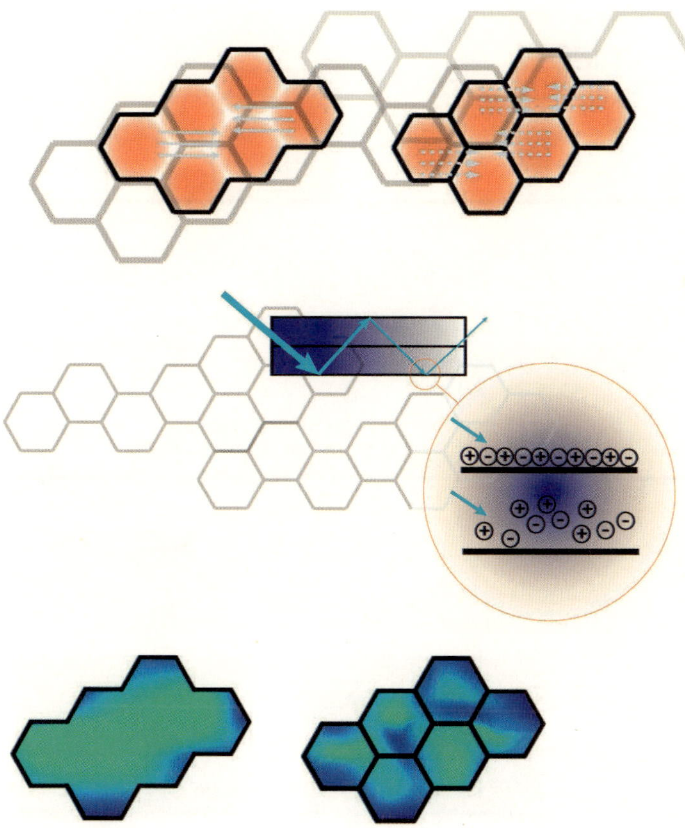

Honeycomb Walls

Opposite Page:
Image of bees and combs

Current Page:
Honeycomb concept diagrams

Student Team:
Li Qian
Yang Zhao
Huadong Zhu

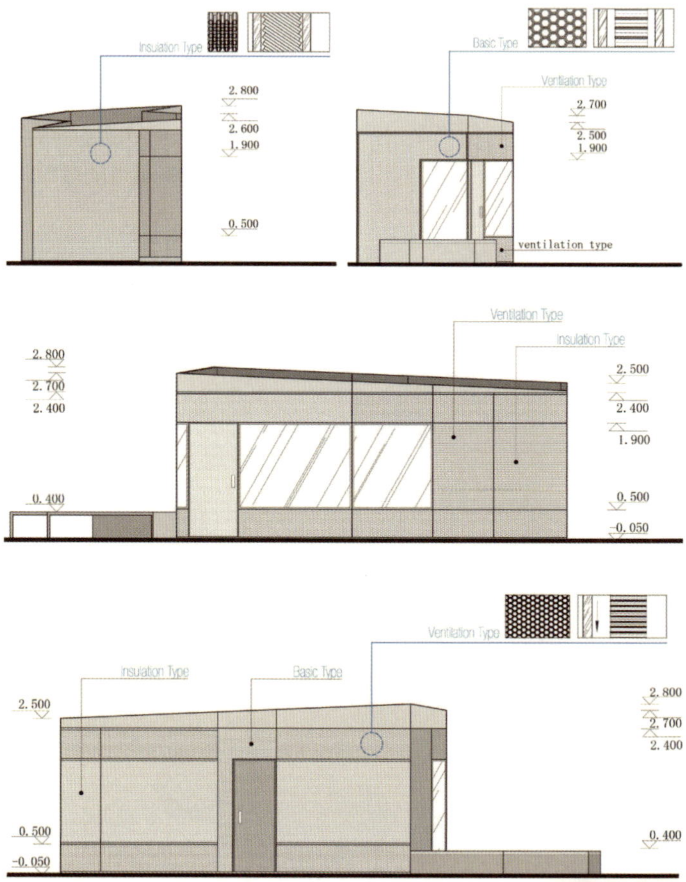

Wall Systems

Top Image:
Guard House Plan

Lower Images:
Guard House Elevations showing location of light reflective wall system

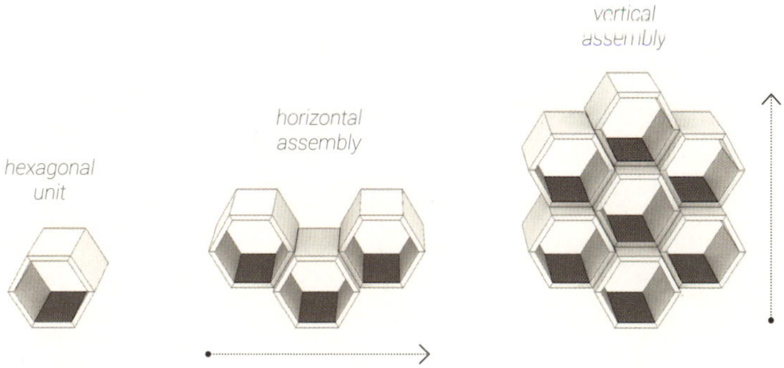

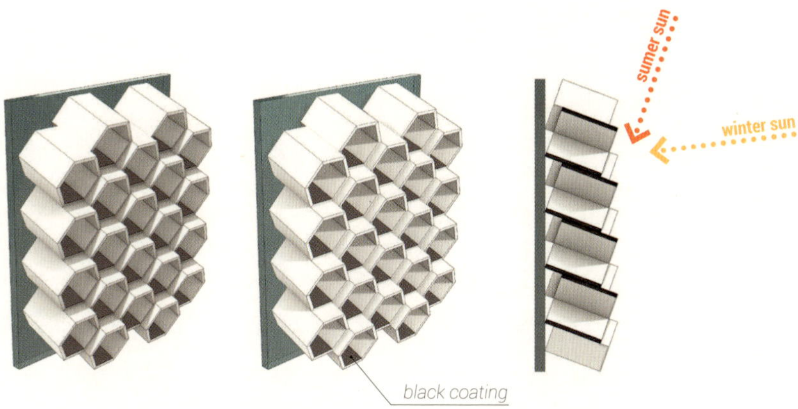

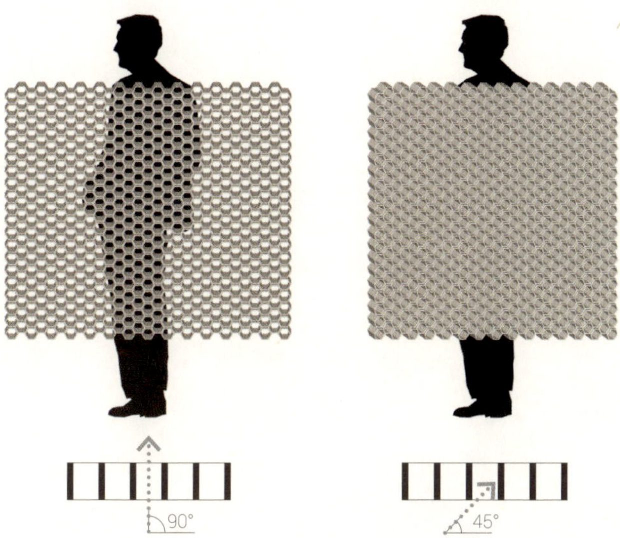

Prototype Studies

Wall System Concept Images showing sunlight reflecting in summer

Wall System Prototype showing summer and winter screening conditions to reflect and absorb light

Microscale Design

In this chapter, the projects looked at nature in detail, at the microscale to understand functions from a new perspective. As at the microscale much of the composition is about fiber orientation, this led the analysis to be concentrated on understanding hierarchies and how change in function occurs at different scales. Rather than select the primary biomimetic organisms such as the lotus leaf, students were tasked with picking any artifact that they were interested in exploring in detail.

Through a series of classes which progressed from studies on form, movement, materiality, and scale, students were asked to develop their work through physical testing, modeling and digital exploration. The outcomes were required to be physically prototyped, which directed the design to tend to be at the industrial design or architectural scale.

The class worked in collaboration with the Department of Molecular and Cell Biology School of Life Science and Technology at Tongji University to help the students develop SEM (Scanning Electron Microscope) images of their artifacts. Professors at this college invited students to use their microscope which used a process of gold coating to allow for magnification. A gold coating was applied to the objects to assist the microscope's electron beam in illuminating the artifacts. This process to study the outcomes directly with the scientists was a unique and one that many times designers don't get to enjoy, as a connection with the sciences can be limited.

The final designs range from proposals for new materials to joints and to systems that provide alternatives to conventional methods such as new concepts for lifting. As the course was open to all students of the university, the composition of students was comprised of undergraduate to graduate from a range of backgrounds such as industrial design, service design, environmental and communication designers. For many, it was their first course in biomimetics and studying nature to inform abstraction for design.

Seed Dispersal

Opposite Page: Photograph of *Aesculus hippocastanum*, Sapindaceae or commonly called a Horse-chestnut tree or conker tree showing flowering inflorescence in early spring

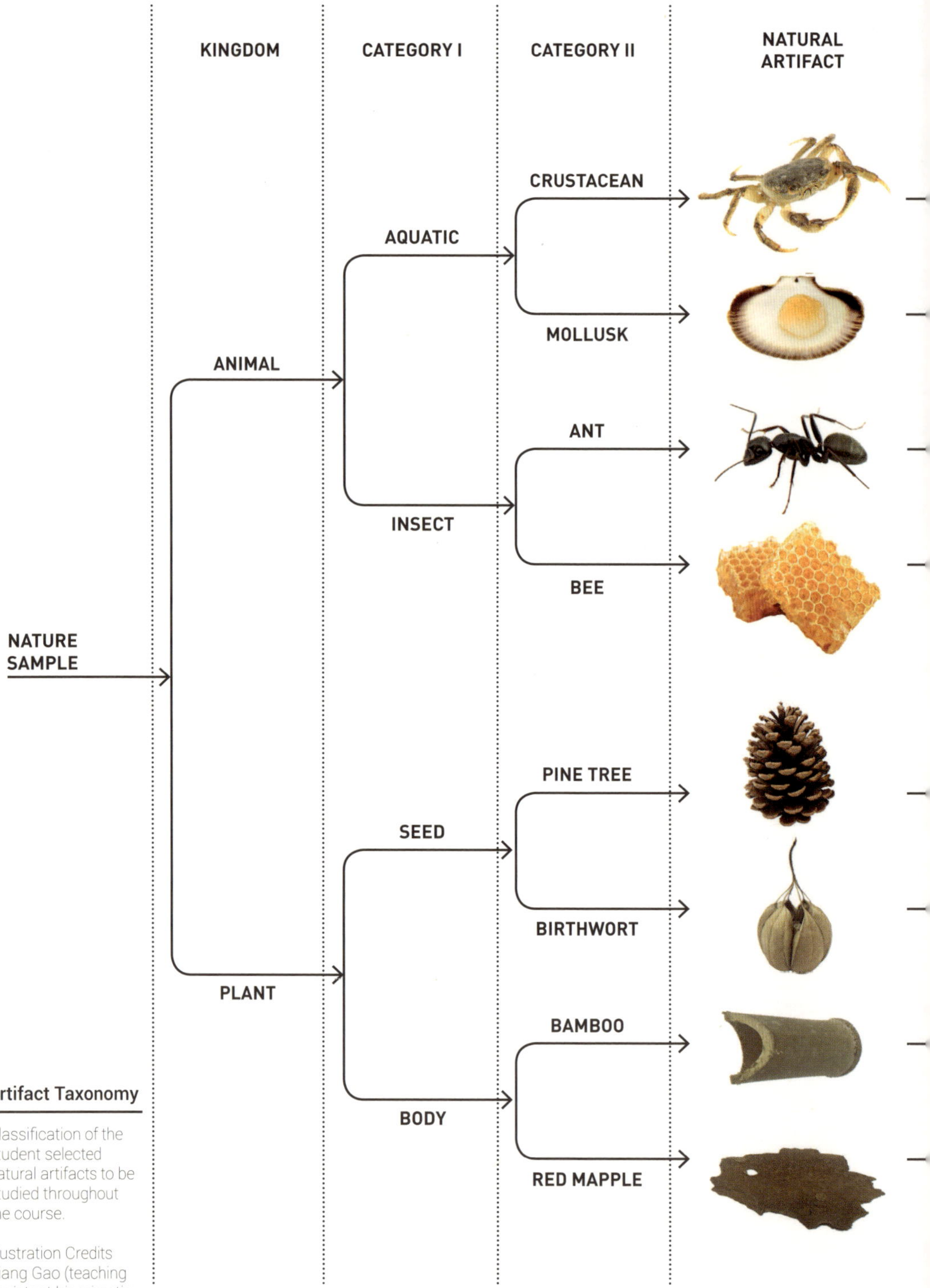

Artifact Taxonomy

Classification of the student selected natural artifacts to be studied throughout the course.

Illustration Credits Xiang Gao (teaching assistant biomimetic design)

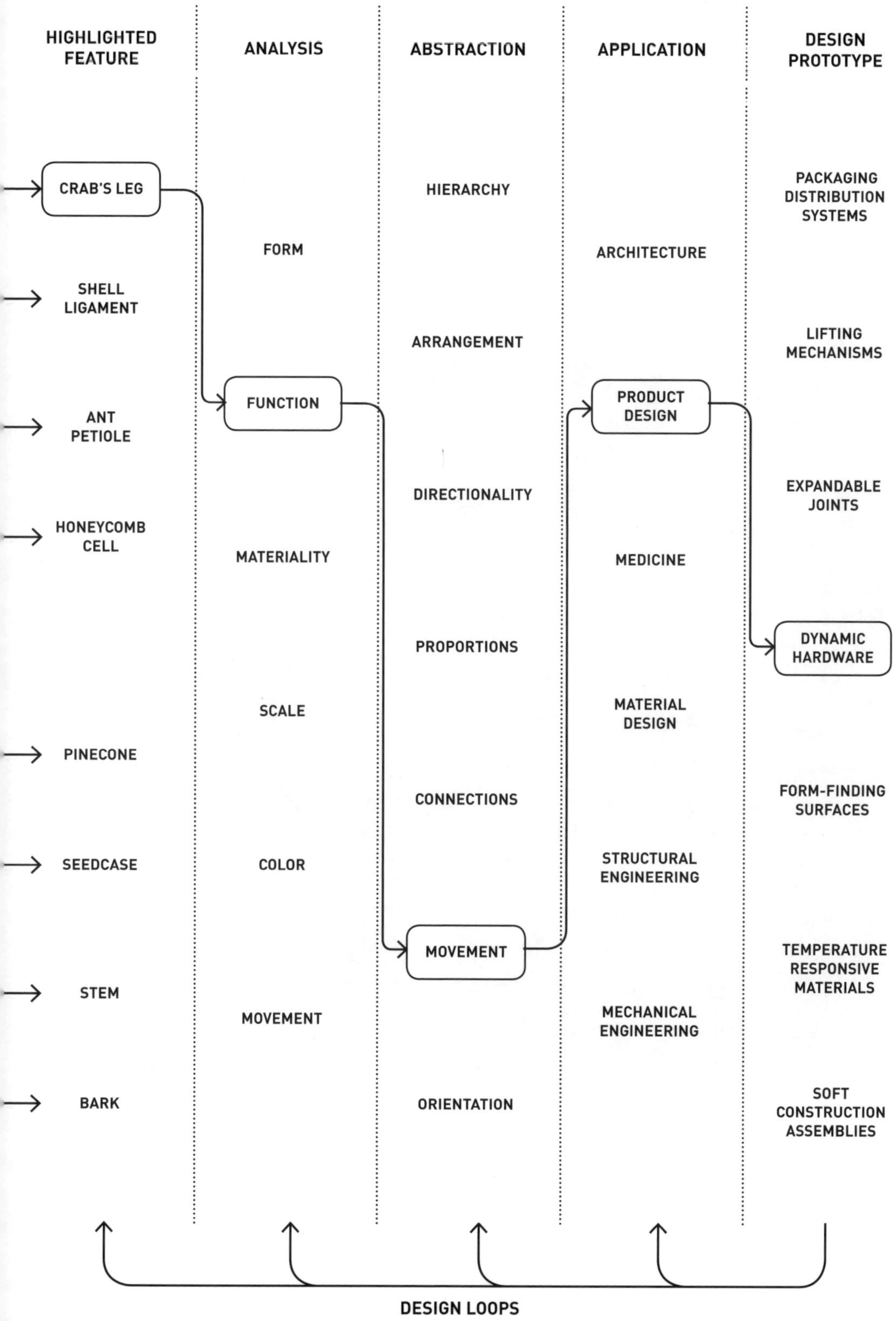
HIGHLIGHTED FEATURE
ANALYSIS
ABSTRACTION
APPLICATION
DESIGN PROTOTYPE
CRAB'S LEG
SHELL LIGAMENT
ANT PETIOLE
HONEYCOMB CELL
PINECONE
SEEDCASE
STEM
BARK
FORM
FUNCTION
MATERIALITY
SCALE
COLOR
MOVEMENT
HIERARCHY
ARRANGEMENT
DIRECTIONALITY
PROPORTIONS
CONNECTIONS
MOVEMENT
ORIENTATION
ARCHITECTURE
PRODUCT DESIGN
MEDICINE
MATERIAL DESIGN
STRUCTURAL ENGINEERING
MECHANICAL ENGINEERING
PACKAGING DISTRIBUTION SYSTEMS
LIFTING MECHANISMS
EXPANDABLE JOINTS
DYNAMIC HARDWARE
FORM-FINDING SURFACES
TEMPERATURE RESPONSIVE MATERIALS
SOFT CONSTRUCTION ASSEMBLIES
DESIGN LOOPS

Hinge Ligament

Shell Ligament

Shells of mollusks have been known for many useful features in the world of biomimetic design, such as re-informing how we build materials at room temperature, or how to build bricks with stronger density compositions. However, in this study, we focused on the feature that allows the shells to open and close repeatedly throughout the life of the mollusk. Typically, the muscles of the mollusk control much of the shells ability to open and close. But, the connection between the two shells that allows for this repetitive movement was the point of interest.

After researching further details on the hinge, known as the ligament, it was found that the material varies from the interior to exterior. The ligament is comprised of two types, one is called the hinge line and is on the exterior of the shell, and the other is the inner ligament called the resilium. These two elements are built to handle the system of compression and expansion which supports the shell and what became the inspiration for the prototype.

This conceptual design explores the idea of a double-layered system which would allow for states of expansion or fully stretched conditions, to a resting state or compressed. The form was chosen for its packing ability and modularity for growth. The properties of the alternating materials would be oriented in different directions to allow for distribution of force throughout the module. This concept would potentially be useful in reconsidering hardware, connections and joints in design, buildings and fashion.

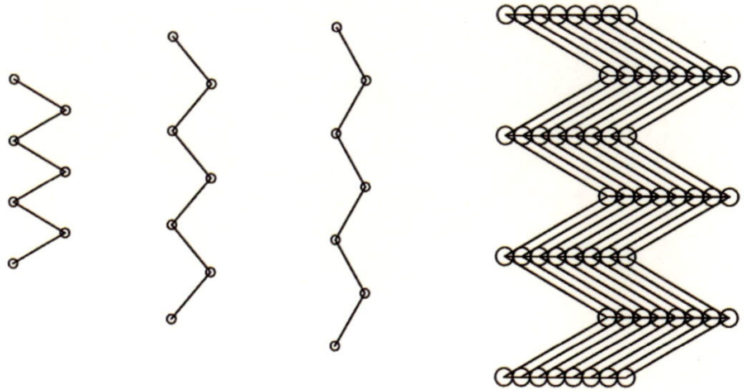

Alternating Hinge

Opposite Page:
Image of mollusk shell

Current Page:
Diagrams showing concept of hinge and alternating modules

Project by
Wanting Tang

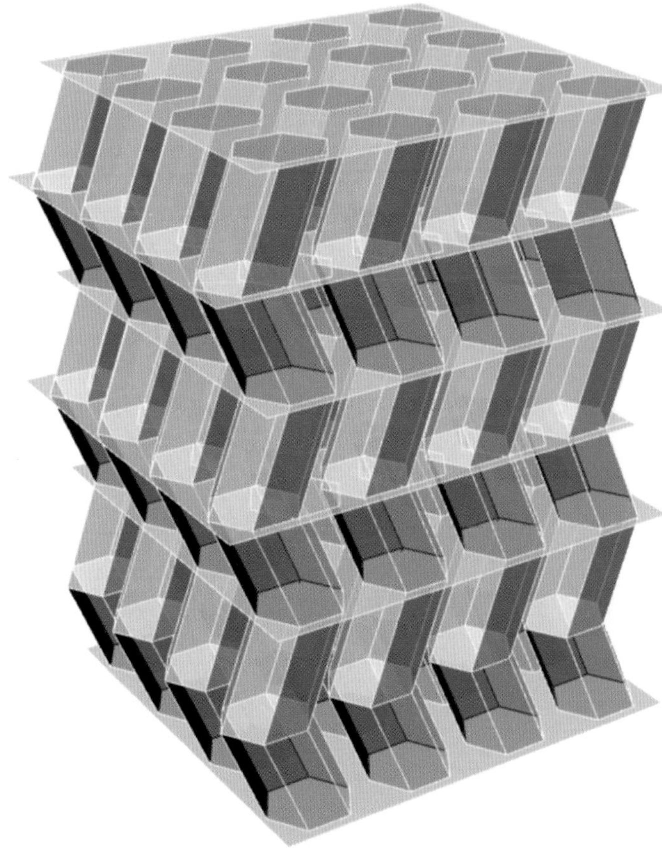

Digital Module

Current Page:
Digital model showing the concept of alternating stacking cells

Opposite Page:
Diagrams showing development concepts of growth and orientation of modules

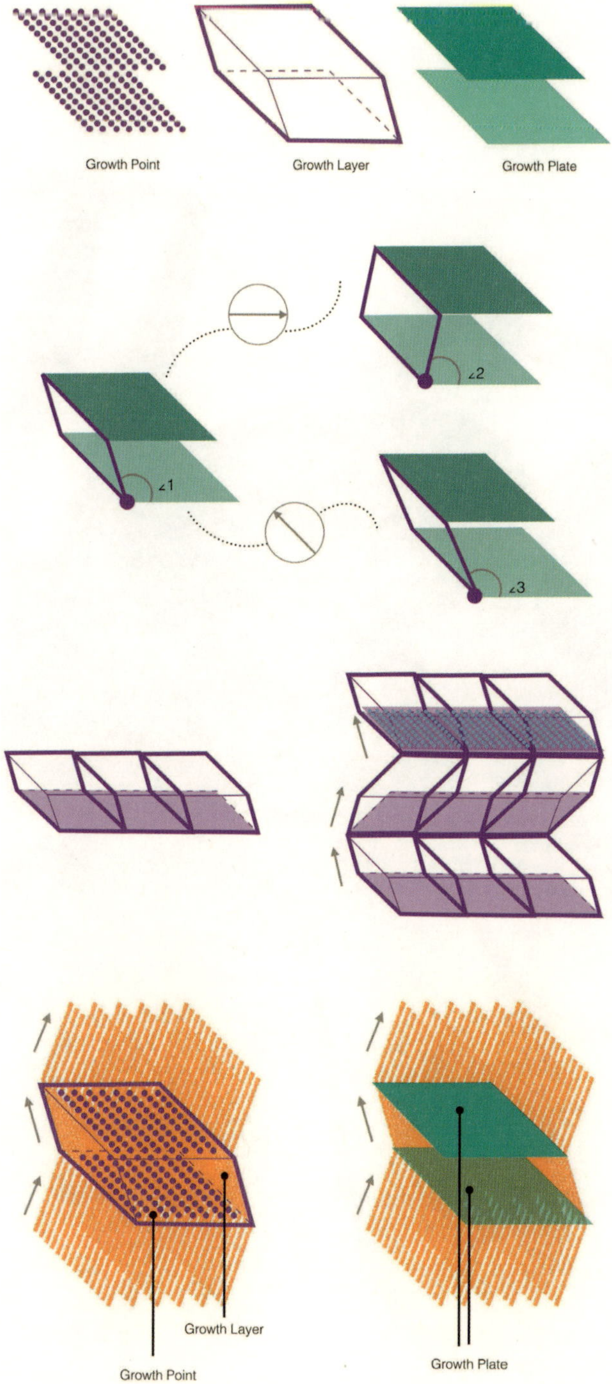
Growth Point
Growth Layer
Growth Plate
∠1
∠2
∠3
Growth Layer
Growth Point
Growth Plate

Ant Petiole

The main focus of this analysis was on the section between the two body volumes, the head and the abdomen of the ant. Between these areas and the connecting thorax, joints enable the ant to rotate these features almost freely. Although it appears to be a paradox, these areas which need to manage the rotation for such a large volume of the outer body compartments, are within the thinnest points in the ant's anatomy.

The petiole differentiates the ant from other insects. "The petiole (and post-petiole, when present) provides a flexible junction, allowing the ant to bend its gaster forward to sting or spray."[1] The petiole is the physical joint of the whole ant body. Even though it constitutes the thinnest area between the several body compartments, its function requires high flexibility while being stable enough to lift the weight of the attached voluminous body parts. The following design principles were derived from this observation investigating the change between flexible and rigid or stable states and conditions. The developed object can be used in several industries to provide a stable and flexible linear joint. This could be useful in architecture or buildings where flexibility and stability are required at the same time.

Joint Concepts

Opposite Page
Photograph of carpenter ant

Current Page:
Photograph with isolation around the petiole

Project by
Fabian Kragenings

1 "Face to Face with an Ant Seeing Red." Scribol.com. May 18, 2018.

Abstracted Joints

The diagrammed mechanism shows the possibility of controlling the angle between two compartments to maintain the flexibility of the joint.

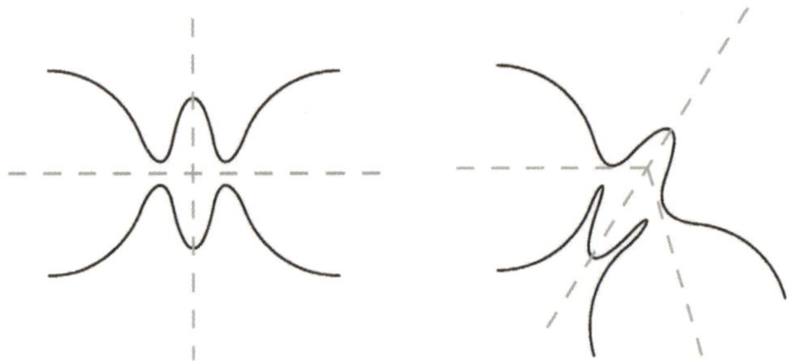

Bending Concepts 1

The abstraction of the mechanism considers the restriction of the angle by the petiole. As shown in the series, there is a gap between the outer volumes and the connecting petiole. This gap defines the rotational range of the compartments.

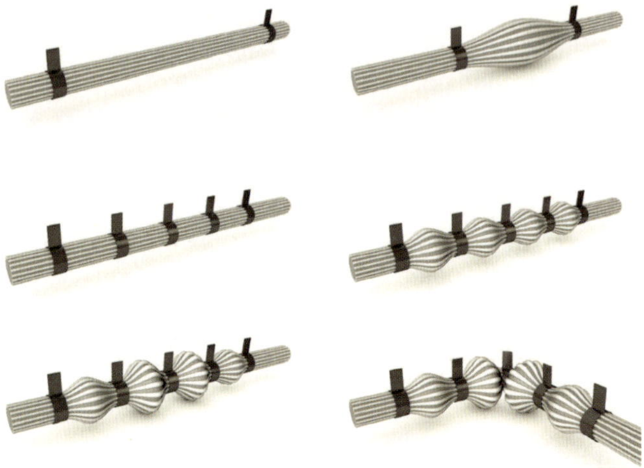

Bending Concepts 2

The concepts to the right explores skin alternating techniques to achieve flexibility and rotation.

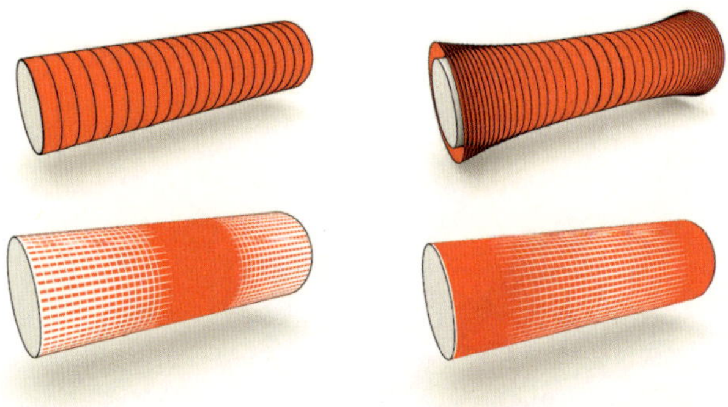

Bending Concepts 3

These studies explore paneling systems that could react to curvature through angled scales to increase or decrease rigidity,

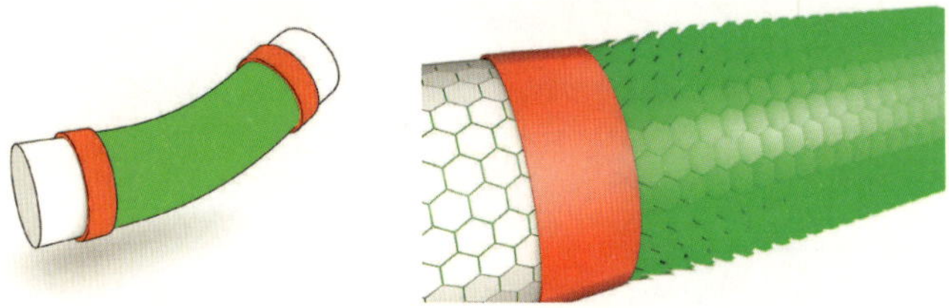

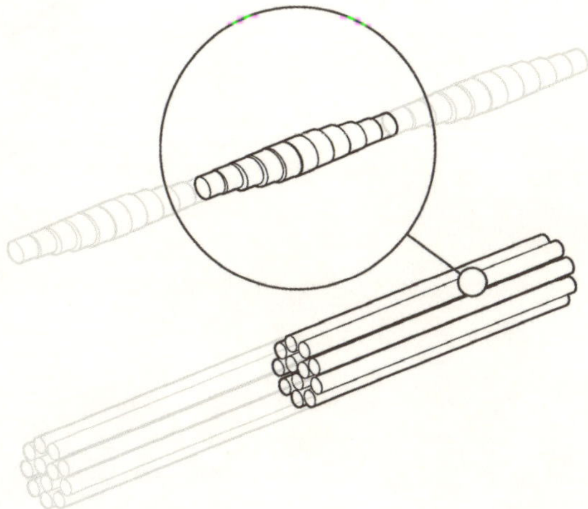

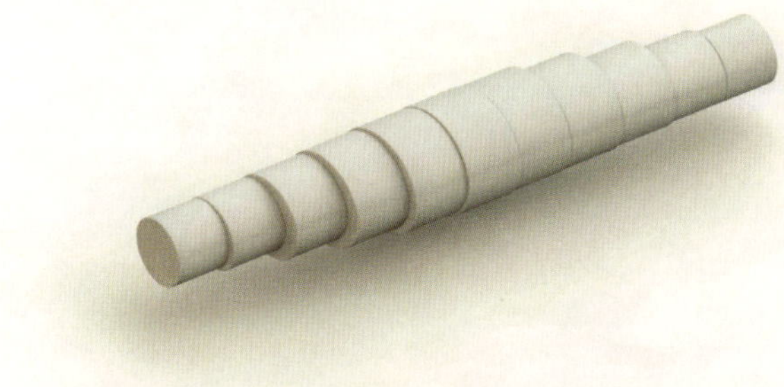

Joint Prototypes

Bundling joint diagram shown at two scales, object and materiality

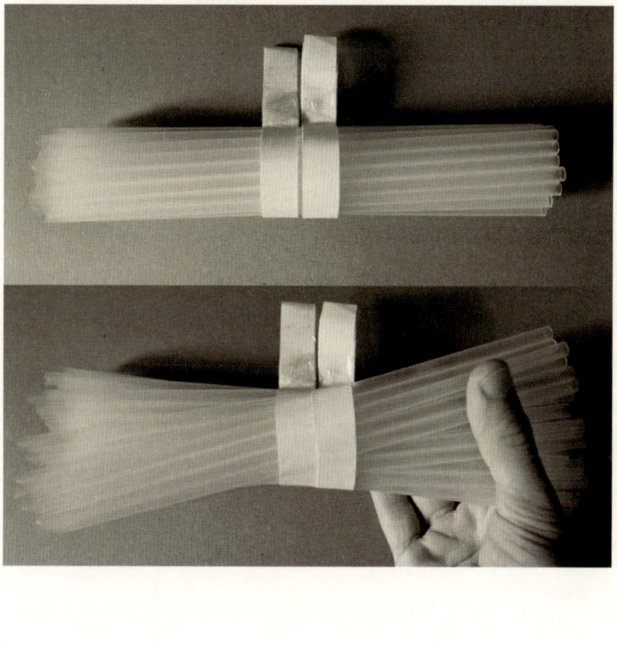

Physical Prototypes

Fiber twisting, due to the holder positioned in the center of the bundle, the individual straws had the ability to move in a radius. By twisting the bundle the overall surface on the outer areas enlarges, which could provide connection points for other objects.

Form Finding

Honeycomb Cells

Honeycomb structures found in beehives are formed through organic material being placed in specific geometric relationships to the bodies of bees and wasps. But another formation occurs on a subtle level in these cells through the form finding the process of all of the other cells working together with forces. Together, the honeycombs form a system that act as a field of tension across the hive.

The concept of form found cellular structures was the interest of this study. An additional level of complexity was added when the development of the forms needed to be movable to adapt to its surroundings, a key aspect found in honeycombs. Through physical prototypes and digital exploration, modules were developed that suited these two directions.

The direction of this work realized that the initial relationship between the support surface and the cells was the most important in ensuring an even distribution of modules. For the prototype development, the challenge was translating a form finding process in terms of materiality and geometry. The end design is a compromise between these two considerations. The end application for this exploration could be useful to packaging, where stabilization of individual compartments is critical.

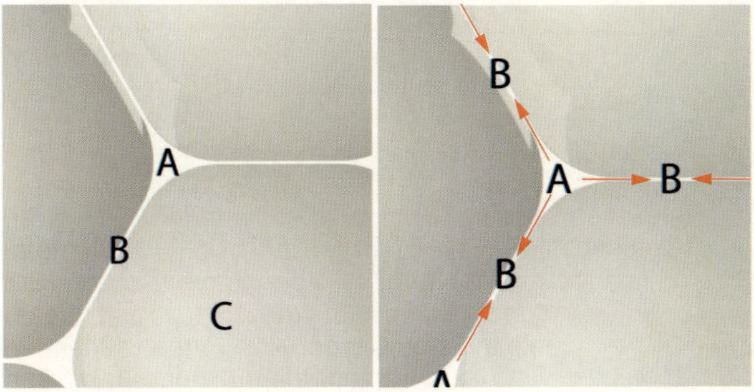

Honeycomb Structures

Opposite Page:
Image
Bees on hive

Current Page:
Diagrams showing
A: Joint
B: 'Border'
C: Liquid Film.
Due to the difference between pressures of Pint A, B and C, the liquid moves to B from A which also create the tension, making the surface flat.

Project by
Qian Li

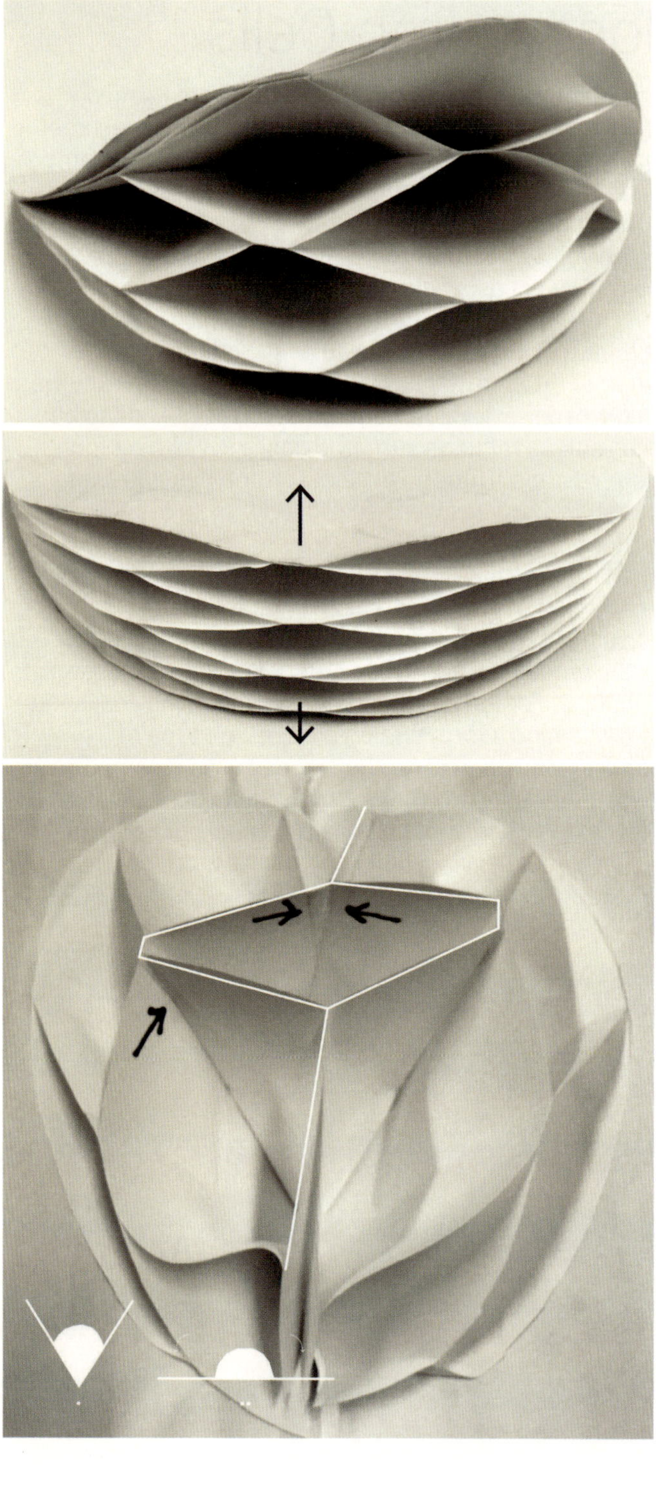

Malleable Prototypes

Current Page:
Photos of flexible cell studies that could allow for form found modular systems

Opposite Page:
Prototypes of soft cells and their compounding packing abilities in relation to their point of support

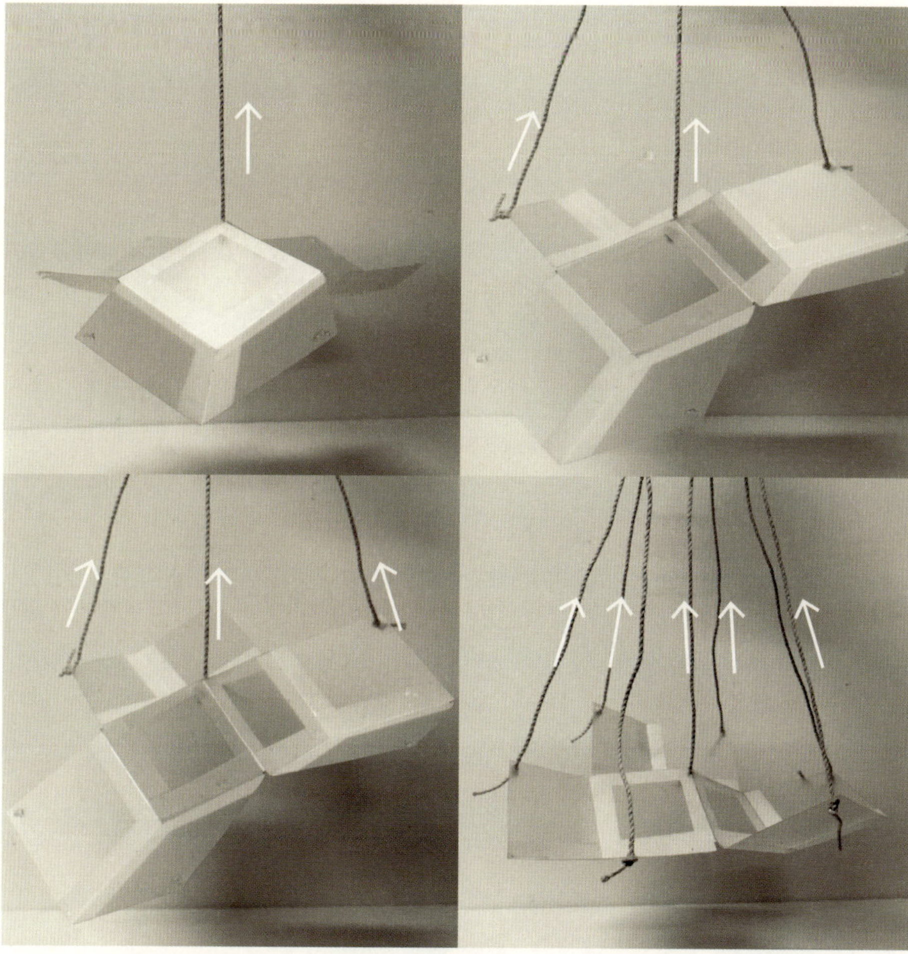

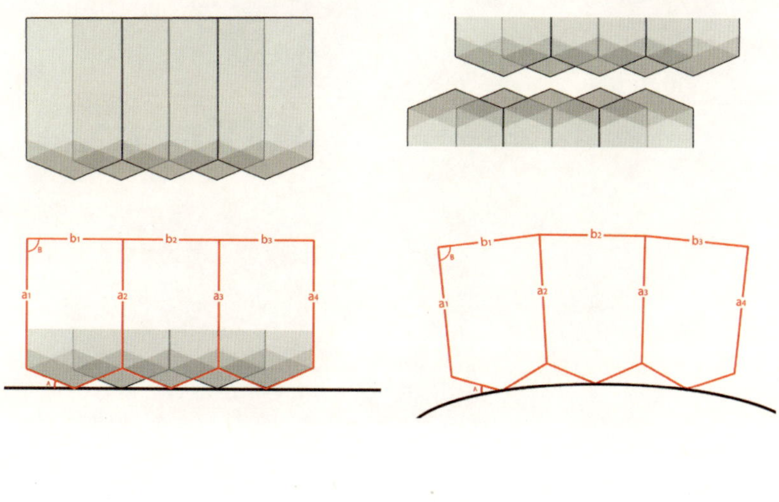
b1
b2
b3
a1
a2
a3
a4
b1
b2
b3
a1
a2
a3
a4

Birthwort Seedcase

The seed case of a birthwort plant was studied for its ability to allow for movement in all directions. This movement is a feature that benefits the seed dispersal abilities once the plant has matured. The strength of the seed case to have this movement is due to an anisotropic joint at one end. At this point, from SEM images, we were able to observe the fiber orientation and abstract the patterns for prototyping.

The prototype explored how to create movement in all directions while opening and closing. The exploration was carried through physical and digital models that examined the material properties necessary to allow for movement. Measurements and ratios were gathered from the natural artifacts to inform the model.

Suitable applications include the packaging industry which could benefit from parcel containers that could be opened and closed infinity in many directions. As well as release mechanisms for emergency objects where quick deployment is necessary.

Seedcase Patterns

Opposite Page:
Photograph of seed case

Current Page:
Diagram showing capabilities of seed case movement

Project By
Laura Raab

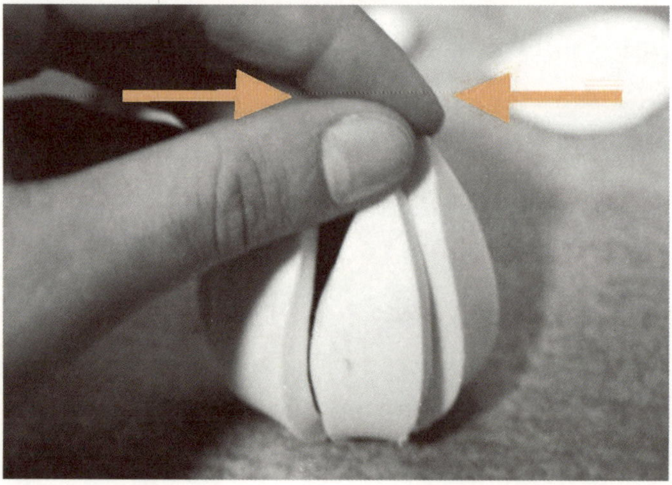

Prototypes

Current Page:
Images of material variation in model

Opposite Page:
Digital developments of seedcase

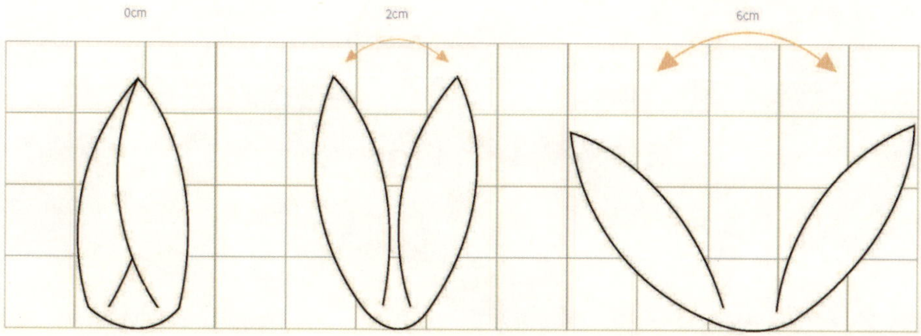
0cm
2cm
6cm

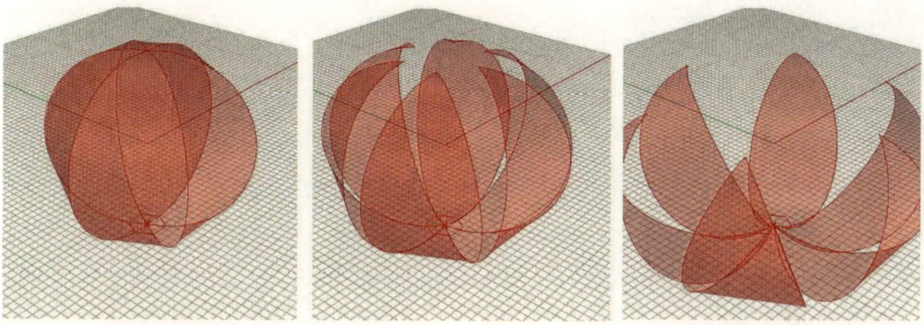

Bamboo

Bamboo is a very force resistant material, but like every material, there is the possibility for breakage. When studying damage in bamboo, one of the most interesting aspects was the way in which cracks are distributed and how these lines of stress follow patterns at a cellular level. From SEM (scanning electronic microscopic) studies, the crack path was studied at various scales. It was noted that the line of force was distributed through the paths of built materials around the fibers and broke open only once a change in scale in the cells was too high.

For this study, we considered how to develop a material that would have predetermined areas for cracking. Applications for this type of study would be suitable for continuous surfaces, such as sidewalks or streets, where pre-programmed regions for breakage could be designed once broken for new programs such as plantings. This concept was explored through physical study and digital exploration to further understand the relationship between the material properties, the scale and arrangement.

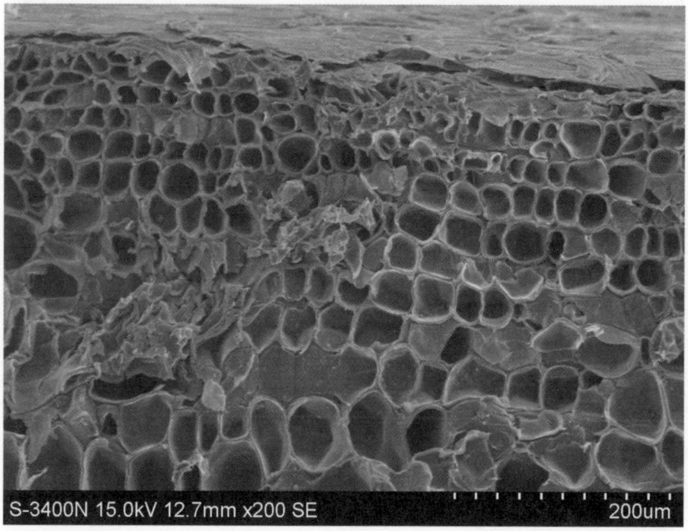

Controlled Crack

Opposite Page:
Photograph of bamboo sections

Current Page:
SEM image of a crack in bamboo.

Project By
Luigi Galbusera

Mock-Up Study

Showing preparation of plaster and bamboo testing for crack allocation. Testing with materials proved hard to use as a control base, therefore digital exploration was considered as a next step to explore relationships between ratios of scale.

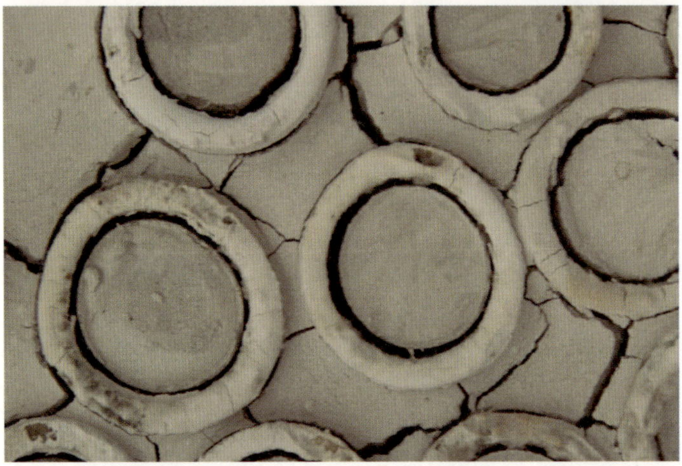

Material Prototyping

Digital model in detail of material allocation

Opposite Page:
Top Image-
Material testing prototypes after breaking

Lower Image-
Digital model showing concept of controlled breaking along lines of attraction

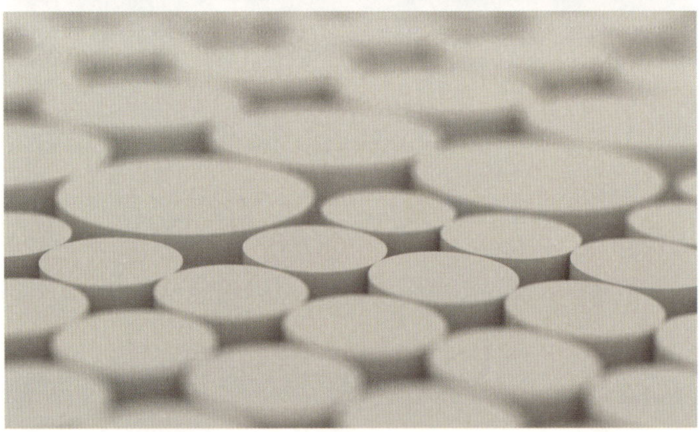

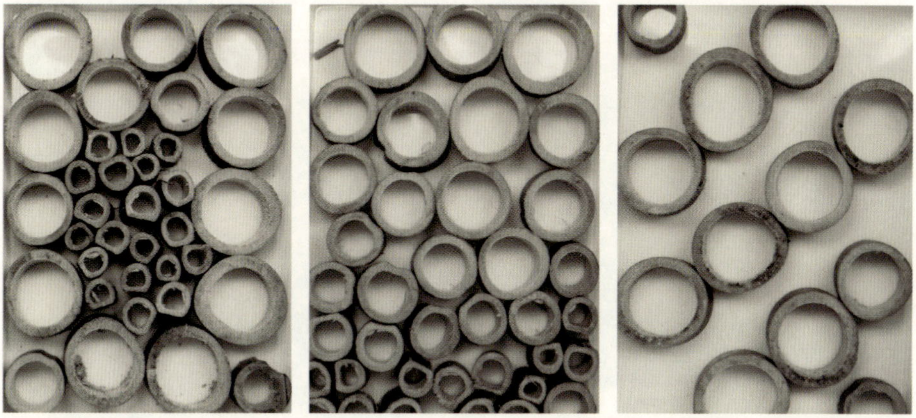

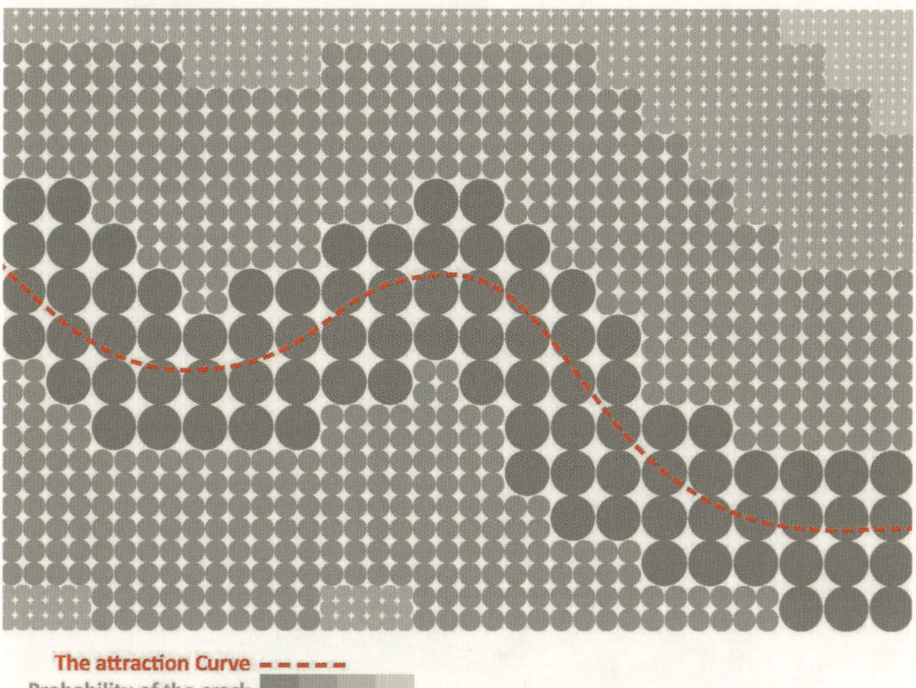
The attraction Curve
Probability of the crack

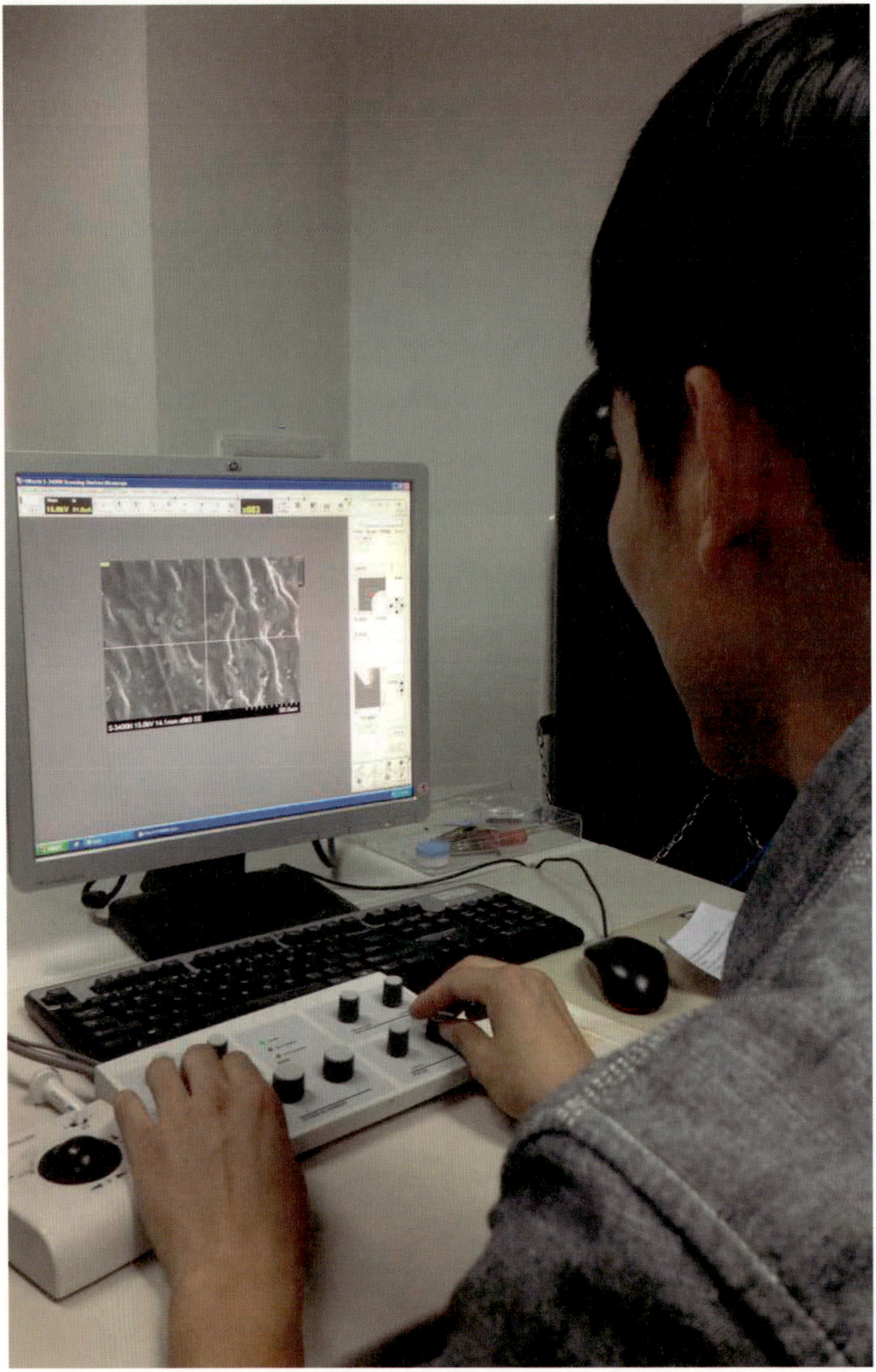

SEM
(Scanning Electron Microscope)

For study at the micron level, students worked with the lab in the Department of Molecular and Cell Biology School of Life Science and Technology at Tongji University to develop imagery exploring the structures. The team in the lab consisted of professor Jian Zhu, Dr Xiling Du and professor Liu Zhu who graciously assisted students in preparing their samples, discussing how to view them to obtain the best results and how to understand the imagery capture. While this process might be standard in the science field, for the design students, it was a rarity to be able to prepare artifacts for scanning and work one on one with the lab.

When we discuss micro and detailed levels of exploration, we are at the scale of a micron (symbol μ) pronounced "mu" and is understood to be 10^{-6}, one-millionth of a meter. A micron is also known to be indicated as a micrometer (μm) which is the same scale and can be represented as um. On many of the students' images, the *um* denotation is indicated for reference. Many of the projects from other labs in this book worked at this scale, as it is where biology and materials begin to show details of their organizational structures. To provide a point of reference, a human hair is about 50 μm in diameter. For measurements below a micron, we commonly use nanometer (nm), which is 10^{-9} one-billionth of a meter.

SEM

Opposite Page: Professor Zhi Liu assisting students with scanning in the Department of Molecular and Cell Biology School of Life Science and Technology. The electron microscope is useful to view smaller details on the surface of objects as it uses accelerated electrons to illuminate details.

Current Page: Image left shows a prepared sample before scanning with artifacts cut down and on adhesive tape. Image right shows samples after gold plating has occurred.

Artifact Prep

A sample being placed on disc for scanning.

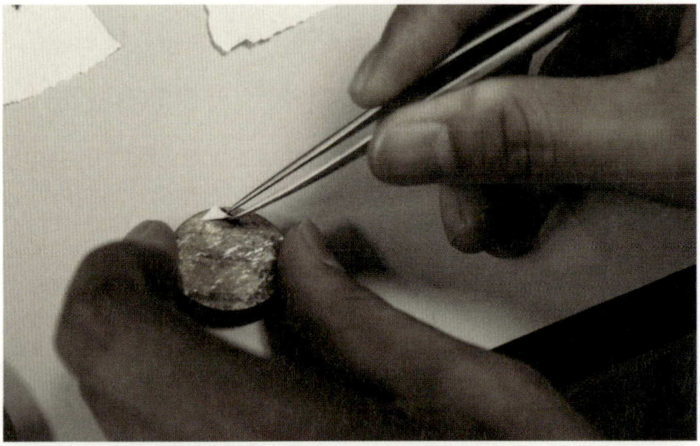

Artifact Prep

Splutter coating chamber where samples are coated before magnification

Artifact Prep

Opened sample chamber of electron microscope

Photographs by Luigi Galbusera

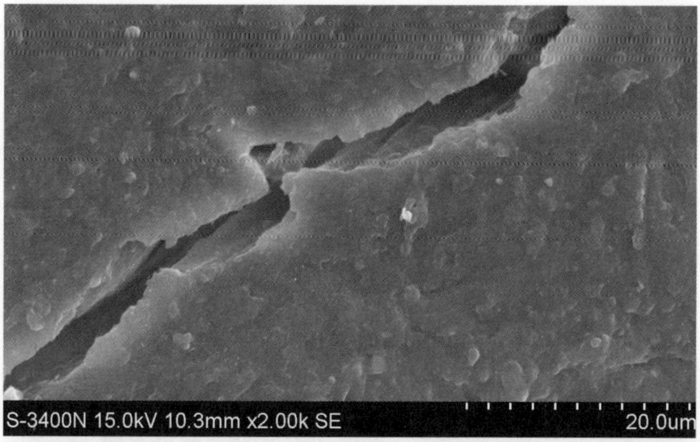

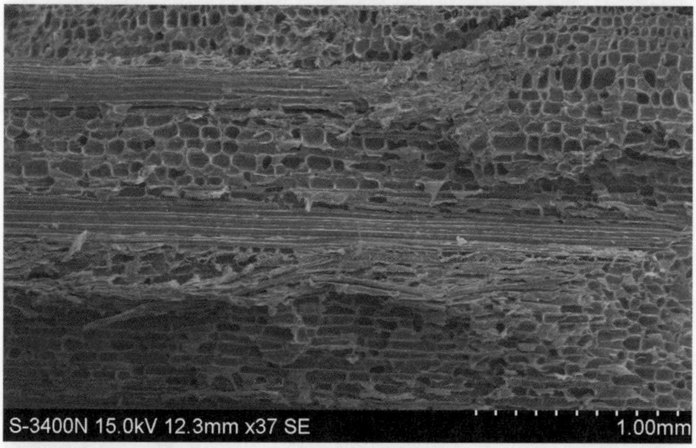

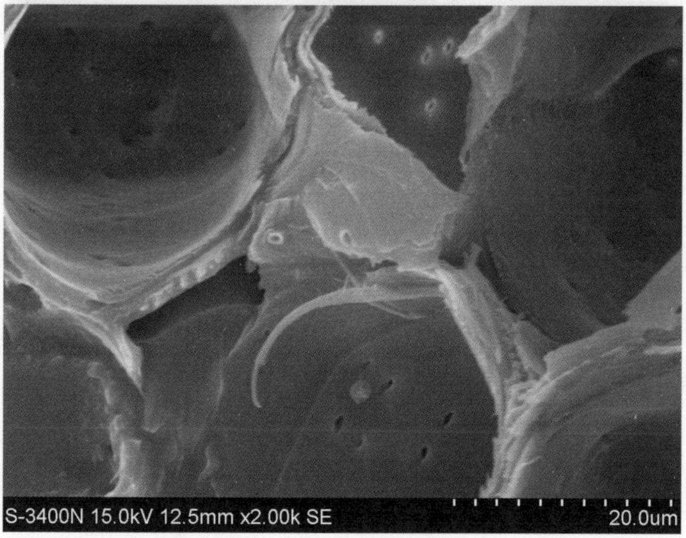

Bamboo Scans

SEM images showing various scales of bamboo. As the project was exploring cracks in the bamboo material, the objective was to investigate the damage of cracks in the tissue.

Photographs by Luigi Galbusera

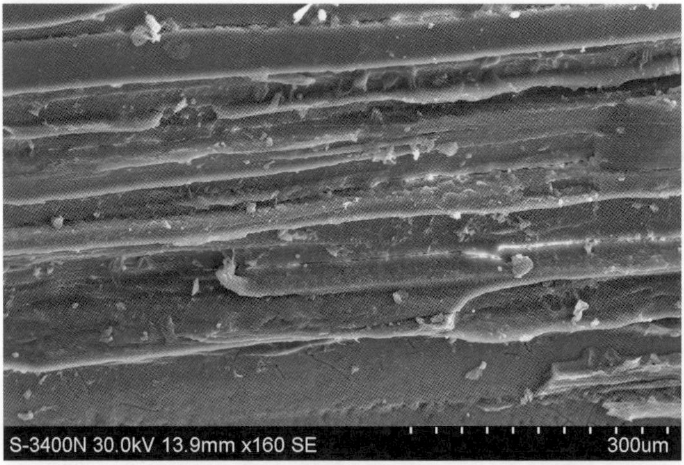

Artifact Scans

Image Top:
Shell outer surface

Image Middle:
Pine cone scale outer surface

Lower Image:
Bamboo end after being broken off

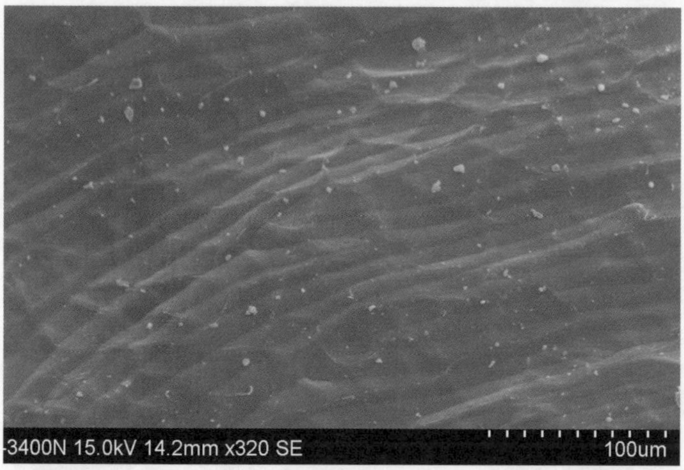

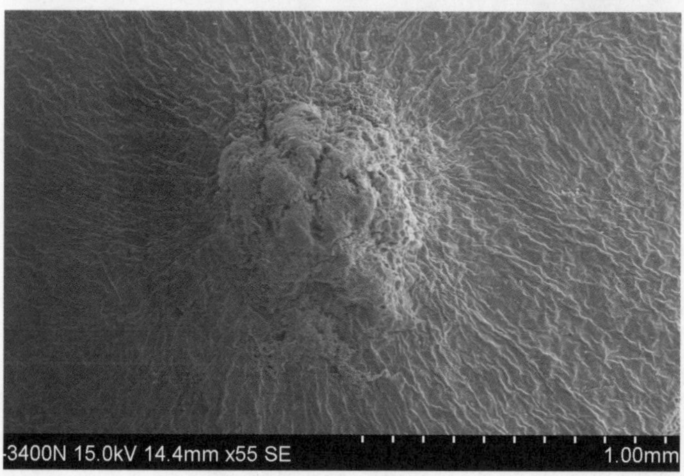

Birthwort Scans

Image Top:
Break in seed case

Image Middle:
Inner surface of seed case

Lower Image:
Main connection of all seedcase shells. Connection was located on the bottom exterior of the seed case.

Conclusion

Tendrils

Fern tendril in early spring

The Potential for Bio-Design in China

What China has achieved in terms of development in the last 30 years has no precedents in history. Averaging an annual 10 percent GDP growth and lifting more than 800 million people out of extreme poverty in such a short time is an undeniable success.[1] Yet it came at a cost: On the one hand, China is today´s world largest emitter of carbon dioxide.[2] On the other, the dramatic urbanization process is resulting in obvious ecological and environmental challenges, including agricultural land loss, polluted rivers, decreased soil quality, endangered species and biodiversity decline.

It is also true that addressing this situation is already part of the government´s current agenda, being this one of the main reasons for the ongoing support of environmental research, green technologies and sustainability. For the last decade, many public and private institutions have been developing their programs to focus on innovation based on nature-inspired solutions. For example, there are more than 20 biomimetic labs established in China, focusing on a wide range of research fields, from health and medicine to physical and applied sciences. However, in these cases, the "bio-" label does not imply a search for innovation in sustainability, but rather an innovation in science and technology. Although such ambitious programs could eventually lead to life-changing breakthroughs, it is also true that such practices sometimes work in isolation and are disconnected from our social and environmental reality, unfortunately missing the "big picture" in the name of progress.

In this regard, biomimicry seems to provide some added value, by merging sustainability and innovation, and

1 "Overview." World Bank. Accessed December 06, 2018. http://www.worldbank.org/en/country/china/overview.

2 Rapier, Robert. "China Emits More Carbon Dioxide Than The U.S. and EU Combined." Forbes. July 01, 2018. Accessed December 06, 2018. https://www.forbes.com/sites/rrapier/2018/07/01/china-emits-more-carbon-dioxide-than-the-u-s-and-eu-combined/#7d437c29628c.

emphasizing the need for a more ethical and holistic understanding of nature, and thereby of the challenges China is facing today. Although both fields use nature as a source, the biomimicry movement goes one step further claiming for an ecological approach that reconnects society with nature.

And here is where the real potential resides as far as education is concerned. Any discipline that involves the study of nature is valuable in its own right, as usually the understanding of a specific subject leads us to respect it more. Imagine then young Chinese design students in a biomimicry class, hearing for the first time about the concept of ecological footprint (everything we do has an impact on natural resources). Or about the carbon footprint (everything we do translates into an emission of carbon dioxide into the atmosphere). Or about the interdependence of organisms in ecosystems (how the loss of just one specie can trigger the extinction of the rest).

The whole Chinese education landscape is currently under a major shift with new universities and high schools adopting STEAM (Science, Technology, Engineering, Arts and Math) and STEM programs. This is exposing younger generations to fields that combine science and design, bringing their attention back to concepts that not so long ago were not as popular, such as creativity or experimentation. The potential impact of this trend in the future of the country cannot be understated.

With these developments and China's push towards becoming an environmental leader, bioinspired design is likely to have a bright future, but technological innovation shouldn't be the only driver. The possibility of a nature-centered education and holistic design agendas in a country with 1.3 billion people is something that literally the whole planet would benefit from. Simply an unmissable opportunity.

Appendix

White Scales

Image of a butterfly wing in detail

Wrinkled Peach

The "wrinkled peach" fungus, species Rhodotus palmatus

Biomimicry Resources

(listed alphabetically)

AskNature - Biomimicry Institute	www.asknature.org
ASU Biomimicry Resource Center	www.biomimicry.asu.edu/resources
B3.8	www.biomimicry.net
Better by Design - An Innovation Guide-Using Natural Design Solutions	www.pca.state.mn.us/sites/default/files/betterbydesign.pdf
Bioinspiration & Biomimetics	www.iopscience.iop.org/journal/1748-3190
Bio-Inspired Design (BID) Community	www.bioinspired.sinet.ca
Biokon	www.biokon.de
Biomimetics from Bath	www.biomimetics.org.uk
Biomimicry Education Network	www.asknature.org/groups/biomimicry-education-network/
Biomimicry Europa	www.biomimicry.eu
Biomimicry examples, tips, & resources - SYNAPSE by Biomimicry 3.8	www.synapse.bio
Biomimicry for Social Innovation	www.bio-sis.net/resources
Biomimicry Institute | Green Schools National Network	www.greenschoolsnationalnetwork.org/services/pdcollaborative/biomimicry-institute
Biomimicry La	www.biomimicryla.org/resources
Biomimicry New England	www.biomimicryne.org/resources
Biomimicry Resource Cards - B 3.8	www.net/the-buzz/resources/biomimicry-resource-cards
Biomimicry Resource Handbook: A Seed Bank of Best Practices	www.biomimicry.net/the-buzz/resources/biomimicry-resource-handbook/volum
Biomimicry Resources - NYSERDA	www.nyserda.ny.gov/Researchers-and-Policymakers/Biomimicry/Biomimicry-Resources
Biomimicry: Designing to Model Nature | WBDG Whole Building Design Guide	www.wbdg.org/resources/biomimicry-designing-model-nature
Bionica Innovatie en Expertise Centrum (Biomimetic Innovation and Expertise Center)	www.bionicacentrum.nl
CEEBIOS (Centre Europeen d'Execellence en Biomimetisme de Senlis)	www.en.ceebios.com
Center for Biologically Inspired Design (CBID)	https://bioinspired.sinet.ca/content/bio-inspired-design-landscape
Da Vinci Global Index (Fermanian Business & Economic Institute)	www.pointloma.edu/centers-institutes/fermanian-business-economic-institute/da-vinci-index-biomimicry
European Biomimicry Alliance	www.biomimicryalliance.eu
Green Harbors Project	www.umb.edu/ghp/resources/biomimicry
Journal of Bionic Engineering	www.jbe.jlu.edu.cn/EN/article/showBrowseTopList.do
Nature + Design towards a Bio-Civilization	www.biomimeticdesign.wordpress.com
North American Association for Environmental Education	www.naaee.org/eepro/resources/sharing-biomimicry-young-people
OMICS International	www.omicsonline.org/biomimetics-journals-conferences-list.php
San Diego Zoo Global	www.techtoreconnect.org
Second Nature Documentary - BIOMIMICRYSA	www.biomimicrysa.co.za/inspiring-links
The Biomimicry Institute	www.biomimicry.org
The Genius of Biome Report	www.hok.com/thought-leadership/genius-of-biome
Think Biomimicry	www.thinkbiomimicry.com/resources
Wyss Institute for Biologically Inspired Engineering (Harvard University)	www.wyss.harvard.edu

Lab List

(by alphabetical order of countries/regions)

AFRICA		
Biomimicry South Africa	SOUTH AFRICA	Cape Town
ASIA		
The Key Laboratory of Bioinspired Smart Interfacial Science and Technology at Beihang University	CHINA	Beijing
The State Key Laboratory of Natural and Biomimetic Drugs	CHINA	Beijing
Research Group Of Professor Lei Jiang , Institute Of Chemistry Chinese Academy Of Sciences	CHINA	Beijing
Department Of Mechanical Engineering, Tsinghua University	CHINA	Beijing
Beijing Institute Of Technology Bionic Robot Mechanism, Perception and Control Laboratory	CHINA	Beijing
The Key Laboratory of Bionic Engineering at Jilin University	CHINA	Changchun
Bai Research Group Zhejiang University	CHINA	Hangzhou
Northeast Forestry University	CHINA	Harbin
Shandong University	CHINA	Jinan
The Dynamic Technology Lab for Bionic Materials and Structures at Donghua University	CHINA	Shanghai
The Biomimetic Design Lab (BiDL), at Tongji University	CHINA	Shanghai
Dt-Lab Donghua University	CHINA	Shanghai
Department of Power Mechanical Engineering, National Tsinghua University	CHINA	Hsinchu
Institute of Nanoengineering And Microsystems National Tsing Hua University	CHINA	Hsinchu
Tzu Chi University	CHINA	Hualien
Dong Hwa University	CHINA	Hualien
Tunghai University	CHINA	Taichung
Biomimicry Taiwan	CHINA	Taipei
National Taiwan University	CHINA	Taipei
School of Medicine, Taipei Medical University	CHINA	Taipei
Naval Materials Research Laboratory	INDIA	Ambernath, Maharashtra
Biomimicry India	INDIA	Bangalore
Innovation, Design Study and Sustainability Laboratory (Ideas Lab), Centre for Product Design and Manufacturing, Indian Institute of Science,	INDIA	Bangalore
Department of Chemical Engineering, Indian Institute of Technology Bombay, Powai	INDIA	Mumbai
Department of Chemistry Indian Institute of Technology Bombay	INDIA	Mumbai
University College of Engineering, MG University, Thodupuzha, Idukki District, Kerala	INDIA	Kerala
Department of Chemistry, University of Mumbai, Vidyanagari	INDIA	Mumbai
Biomimetics Lab	IRAN	Rasht, Iran
Chitose Institute of Science and Technology	JAPAN	Chitose
Hamamatsu University School of Medicine	JAPAN	Hamamatsu City, Shizuoka

Biomimetic and Technomimetic Molecular Science Laboratory	JAPAN	Nara, Ikoma
Advanced Polymer Materials Lab	JAPAN	Osaka
Tohoku University	JAPAN	Sendai
Tokyo University of Agriculture Faculty	JAPAN	Tokyo
Research Center for Advanced Science And Technology, the University of Tokyo	JAPAN	Tokyo
Yoshioka Lab	JAPAN	Tokyo
National Museum of Nature And Science	JAPAN	Tokyo
Aero/Aqua Biomimetics Lab, Chiba University	JAPAN	Tokyo
Research Center For Advanced Science And Technology, The University Of Tokyo	JAPAN	Tokyo
Yoshioka Lab	JAPAN	Tokyo
National Museum Of Nature And Science	JAPAN	Tokyo
Aero/Aqua Biomimetics Lab, Chiba University	JAPAN	Tokyo
Japan National Institute of Material Science	JAPAN	Tsukuba, Ibaraki
Biomimicry Singapore	SINGAPORE	Singapore
Biological & Biomimetic Materials Laboratory	SINGAPORE	Singapore

CENTRAL AMERICA		
Biomimicry Costa Rica	COSTA RICA	Costa Rica
EUROPE		
Bioversum	AUSTRIA	Eisenstadt
Bio Inspired Engineering at the University of Applied Sciences	AUSTRIA	Kufstein
Fachhochschule Kufstein Tirol	AUSTRIA	Kufstein
Bionik/ Biomimetics in Energy System	AUSTRIA	Villach
Fachhochschule Kärnten Campus Villach	AUSTRIA	Villach
Biomimicry Be	BELGIUM	Brussels
Biomimicry Europa Aisbl	BELGIUM	Brussels
Greenloop Sa	BELGIUM	Brussels
University of Zadar, Biomimicry Livinglab	CROATIA	Zadar
Biommicry Europa	FRANCE	Paris
Ceebios	FRANCE	Senlis
Comite Francais De Biomimicry Europa	FRANCE	Paris
Rwth Bionics Center	GERMANY	Aachen
Bayreuth, University, Biomaterials	GERMANY	Bayreuth
Berlin Technical University, Fg Bionics	GERMANY	Berlin
Berlin, Technical University, Fg Massive Construction	GERMANY	Berlin
Biokon	GERMANY	Berlin
Biomimicry Germany	GERMANY	Berlin
Phi360	GERMANY	Berlin
Steinbeis Sibe: Bionics Management	GERMANY	Berlin
University of Applied Sciences And Bielefeld University	GERMANY	Bielefeld
Bionik	GERMANY	Bocholt
Westphalian Univeristy, Hs Gelsenkirchen	GERMANY	Bocholt
University of Bonn, Nees Institute	GERMANY	Bonn
Bremen University of Applied Sciences	GERMANY	Bremen

EUROPE		
Hochschule Bremen: Biomimetics-Innovation-Centre	GERMANY	Bremen
Technical University Darmstadt, Sla (Department of Fluid Mechanics And Aerodynamics)	GERMANY	Darmstadt
Cluster Bayonik	GERMANY	Deggendorf
The University of Freiburg, Biology	GERMANY	Freiburg
Albert-Ludwigs-Universität Freiburg	GERMANY	Freiburg Im Breisgau
Functional Morphology, Biomechanics and Biomimetics, University of Freiburg	GERMANY	Freiburg Im Breisgau
University of Applied Sciences Hamm-Lippstadt, Material Design - Bionics and Photonics	GERMANY	Hamm
Heidelberg, Fh	GERMANY	Heidelberg
Steinbeis Sibe: Bionics Management	GERMANY	Herrenberg
Technical University Ilmenau, Biomechatronics	GERMANY	Ilmenau
University of Karlsruhe, Bionics Laboratory	GERMANY	Karlsruhe
Technical University of Munich - Leonardo Da Vinci Center	GERMANY	Munich
University of Applied Sciences Rhine-Waal - Bionics / Biomimetics	GERMANY	Rhine
Saarland University of Applied Sciences	GERMANY	Saarbrücken
Steinbeis Sibe: Bionics Management	GERMANY	Stuttgart
Tübingen, University: Medical Technology	GERMANY	Stuttgart
University of Applied Sciences Carinthia	GERMANY	Villach
Phi360	HUNGARY	Budapest
Central European University, Environmental Sciences & Policy Dept.	HUNGARY	Budapest
Biomimicry Italy	ITALY	Trento
Planet/Biomimicry Italy	ITALY	Brescia
Universität Pisa	ITALY	Pisa
Biomimicry Netherlands	NETHER-LANDS	Eindhoven
Biomimicry NL	NETHER-LANDS	Utrecht
University of Edinburgh Scotland	SCOTLAND	Edinburg
Biomimicry Iberia	SPAIN	Granada
Instituto Europeo de Diseno de Madrid	SPAIN	Madrid
Kth – Royal Institute of Technology, Swedish Centre of Bio Mimetic Fibre Engineering (Biomimet) (Sweden)	SWEDEN	Stockholm
Biomimicry Switzerland	SWITZER-LAND	Basel
Bio-Inspired Projects Platform	SWITZER-LAND	Lausanne
University of Bath	UK	Bath
University of Birmingham	UK	Birmingham
University of Bristol	UK	Bristol
University of Cambridge	UK	Cambridge
Imperial College London: Institute of Biomedical Engineering	UK	London
Exploration Architecture	UK	London
Biomimicry Uk	UK	York

NORTH AMERICA		
Biomimicry Alberta	CANADA	Calgary
Biomimicry Quebec	CANADA	Montreal, Quebec
Biomimicry Yucatan	MEXICO	El Cuyo, Yucatan
Biomimika	MEXICO	Mexico City
Biomimicry Mexico	MEXICO	Mexico City
Bric	USA	Akron, Ohio
Georgia Institute of Technology, Centre for Biologically Inspired Design (CBID)	USA	Atlanta, Georgia

Biomimicry Texas	USA	Austin, Texas
University of California At Berkeley, Bioengineering	USA	Berkeley, California
Biomimicry New England	USA	Boston, Massachusetts
Harvard University, School of Engineering and Applied Sciences, the Aizenberg Bio Mineralization and Biomimetics Lab.	USA	Boston, Massachusetts
Biomimicry Chicago	USA	Chicago, Illinois
Great Lakes Biomimicry	USA	Cleveland, Ohio
Biomimicry Colorado	USA	Denver, Colorado
Uc-La	USA	Los Angeles, California
Biomimicry La	USA	Los Angeles, California
Design Studio	USA	Los Angeles, California
Biomimicry Institute	USA	Missoula, Montana
Biomimicry B3.8	USA	Missoula, Montana
Biomimicry NYC	USA	New York City
Biomimicry Northern Forest	USA	New York City
State University of NY, College of Environmental Science and Forestry	USA	New York City
California State University Northridge , Biology	USA	Northridge, California
Biomimicry Oregon	USA	Portland, Oregon
The Nature Lab	USA	Providence, Rhode Island
University of San Diego	USA	San Diego, California
Biomimicry San Diego	USA	San Diego, California
Bay Area Biomimicry Network	USA	San Francisco, California
Coastal Redwood Biomimicry	USA	San Francisco, California
Biomimicry Puget Sound	USA	Seattle, Washington
Liko Lab	USA	Seattle, Washington
Bio-X	USA	Stanford, California
Onondaga Community College	USA	Syracuse, New York
The Biomimicry Center, Arizona State University	USA	Tempe, Arizona
The Biodesign Institute, Arizona State University	USA	Tempe, Arizona

OCEANIA

Auckland Bioengineering Institute	AUSTRALIA	Auckland
Curtin University of Technology	AUSTRALIA	Bentley Western Australia
Biomimicry Australia, Jane Toner	AUSTRALIA	Kew, Victoria
Biomimicry Swarm Australia	AUSTRALIA	Melbourne
Biomimetic Robotics Lab	AUSTRALIA	Queensland
The University of Sydney	AUSTRALIA	Sydney
Institute For Frontier Materials	AUSTRALIA	Victoria
Curtin University of Technology	AUSTRALIA	Bentley Western Australia

SOUTH AMERICA

Biomimicry Colombia	COLOMBIA	Bogota
Biomimicry Argentina	ARGENTINA	Buenos Aires
Biomimicry Caribe	COLOMBIA	Colombia
Biomimicry Chile	CHILE	Santiago
Biomimicry Brasil	BRAZIL	São Paulo

Lab Maps Disclaimer:

Labs shown should not to be taken as all inclusive lists of labs throughout the world. Although extensive research was undertaken, due to language capabilities, and access to on line platforms, there may be labs not indicated in the map.

Index

A

B

C

D

E

F

G

H

I

J

L

M

N

Bibliography

Introduction

Bagley, Rebecca O. 2014. "Biomimicry: How Nature Can Streamline Your Business For Innovation." *Forbes*. April 15, 2014.

Bagley, Rebecca O. 2014. "Five Tech Trends That Can Drive Company Success." *Forbes*. April 1, 2014

Bioinspiration: An Economic Progress Report. San Diego Zoo Global, 2013, Bioinspiration: An Economic Progress Report

China. | Climate Action Tracker. Accessed December 1, 2018. https://climateactiontracker.org/countries/.

Gundlach, Carsten. "Bionics in Patents - Semantic Base Analysis for the Exploitation of Bionic Principles in Patents." In Current Scientific and Industrial Reality: Proceedings of the TRIZ-Future Conference 2007 ; Frankfurt, Germany, November, 6th - 8th, 2007. Kassel: Kassel Univ. Press, 2007.

Harkness, Jon M. "In Appreciation: A Lifetime of Connections: Otto Herbert Schmitt, 1913 - 1998." *Physics in Perspective (PIP)* 4, no. 4 (2002): 456-90. doi:10.1007/s000160200005

Lessons of Design Learned from Nature." 2011. *The Economist*. November 9, 2011.

Murray, Sarah. 2018. "Biomimicry Unleashes Wave of Materials Innovation." *Financial Times*. January 8, 2018.

"Overview." World Bank. Accessed December 06, 2018. http://www.worldbank.org/en/country/china/overview

"Rapier, Robert. "China Emits More Carbon Dioxide Than The U.S. and EU Combined." Forbes. July 01, 2018. Accessed December 06, 2018. https://www.forbes.com/sites/rrapier/2018/07/01/china-emits-more-carbon-dioxide-than-the-u-s-and-eu-combined/#7d437c29628c.

"The Da Vinci Index & Biomimicry." Point Loma Nazarene University, www.pointloma.edu/centers-institutes/fermanian-business-economic-institute/da-vinci-index-biomimicry.

Vincent, Julian F.v., Olga A. Bogatyreva, Nikolaj R. Bogatyrev, Adrian Bowyer, and Anja-Karina Pahl. "Biomimetics: Its Practice and Theory." *Journal of The Royal Society Interface 3*, no. 9 (2006): 471-82. doi:10.1098/rsif.2006.0127.

Yoram, Helfman Cohen Yael. Reich. *Biomimetic Design Method for Innovation And Sustainability*. S.l.: Springer, 2018

Design Education Methods

"Face to Face with an Ant Seeing Red." Scribol.com. May 18, 2018.

Henrion, Wolfgang, and Helmut Tributsch. "Optical Solar Energy Adaptations and Radiative Temperature Control of Green Leaves and Tree Barks." *Solar Energy Materials and Solar Cells* 93, no. 1 (2009): 98-107. doi:10.1016/j.solmat.2008.08.009.

"Houtan Park." *Landscape Voice*. September 22, 2013

"Luxun Park." *Beijing Visitor - China Travel Guide*. December 13, 2017.

"WATER PURIFICATION_ Shanghai Houtan Park: Landscape as a Living System." Encuentura. November 24, 2014.

Natural Artifacts

"Anisotropic." Biology Online. Accessed November 22, 2018. https://www.biology-online.org/dictionary/Anisotropic.

Chinga-Carrasco, Gary. "Cellulose Fibres, Nanofibrils and Microfibrils: The Morphological Sequence of MFC Components from a Plant Physiology and Fibre Technology Point of View." *Nanoscale Research Letters* 6, no. 1 (2011): 417. doi:10.1186/1556-276x-6-417.

"Definition of 'isotropic'." Take Heed/pay Heed Definition and Meaning | Collins English Dictionary. Accessed November 22, 2018. https://www.collinsdictionary.com/dictionary/english/isotropic.

"Hydrophilic." Biology Online. Accessed November 22, 2018. https://www.biology-online.org/dictionary/Hydrophilic

"Hydrophobic." Biology Online. Accessed November 22, 2018. https://www.biology-online.org/dictionary/Hydrophobic.

"Phyllotaxis." From Wolfram MathWorld. Accessed November 22, 2018. http://mathworld.wolfram.com/Phyllotaxis.html.

"Seta," Merriam-Webster, , accessed November 22, 2018, http://www.merriam-webster.com/dictionary/seta.

"Trichome," Biology Online, , accessed November 22, 2018, https://www.biology-online.org/dictionary/Trichome.

Figure Index